The *Field Guide* TO
HORSES

By Daniel and Samantha Johnson

Voyageur Press

First published in 2009 by Voyageur Press, an imprint of MBI Publishing Company, 400 First Avenue North, Suite 300, Minneapolis, MN 55401 USA

Voyageur Press titles are also available at discounts in bulk quantity for industrial or sales-promotional use. For details write to Special Sales Manager at MBI Publishing Company, 400 First Avenue North, Suite 300, Minneapolis, MN 55401 USA.

To find out more about our books, visit us online at www.voyageurpress.com.

Library of Congress Cataloging-in-Publication Data

Johnson, Daniel, 1984–
The field guide to horses / by Daniel and Samantha Johnson.
        p. cm.
Includes index.
ISBN 978-0-7603-3508-6 (flexibound)
1. Horse breeds. 2. Horses—Identification. I. Johnson, Samantha. II. Title.
SF291.J526 2009
636.1—dc22

                              2009014652

ISBN-13: 978-0-7603-3508-6

On the title page: Like mother, like son. This young Oldenburg colt is a very promising individual and possesses the breed's characteristics.
On the contents page: The impressive majesty of horses is fully evident in this portrait of a chestnut Oldenburg mare.

Editor: Amy Glaser

Designed by: Kazuko Collins

Cover designed by: LeAnn Kuhlmann

Printed in China

Contents

Acknowledgments 5

Chapter 1
An Introduction to Horses 6

Chapter 2
Beyond the Basics 14

Chapter 3
Identifying Horse Colors 18

Chapter 4
White Markings and Other Identifying Characteristics 32

Chapter 5
Coat Patterns 42

Chapter 6
Breed Profiles 54

Glossary 143

Index 144

Dedication

To LJ and PJ—with love

Acknowledgments

We wish to recognize with gratitude the following individuals who assisted with this project:

- Our editor, Amy Glaser, for her continued support and encouragement!
- Jayme Van Haverbeke-Nelson and everybody at PineRidge Equestrian Center, for always being there for us and for making it fun! Thanks!
- Jim, Reita, and Tasha Gelander, for helping to arrange photography shoots and for providing wonderful equine models.
- Jesica Retzleff, for allowing us to photograph the handsome cover boy.
- Miracle Welsh Mountain Ponies, for the use of their facility and pony models.
- J. Keeler Johnson, for the use of his exceptional draft horse photography—you're great!
- Emily and Anna Johnson, for always lending a helping hand when we need it!
- Lorin Johnson for proofreading and Paulette Johnson for photo editing—thank you so much!
- Cadi, just because.
- Herbie and Junior, just because, too.

Special acknowledgment goes to the following photographers who graciously allowed their images to be reproduced: Cheryl Gibson (New Forest Pony), Connie Summers (Clydesdale, Shagya Arabian, Friesian, Colonial Spanish Mustang), Dorthy Brown (Exmoor), Dru Harper (Caspian), Marilou Seabrook (Lac La Croix Indian Pony), Rita Kane (Gotland), Sheila McKinnon (Eriskay), Marcia Boezeman (Canadian), Kirsten Barry (German Riding Pony), Wendy Bridges (Highland, Dartmoor), and Panache Hackney Horse Farm (Hackney Horse). Also special thanks to Jane Mullen, Vicki Hudgins, Joyce Covington, Leslie Bebensee, and Brauns Ridge Farm, for assisting us with this project—your help is greatly appreciated!

Opposite: There's nothing cuter than a newborn foal! This chestnut Welsh Mountain Pony filly is only two weeks old. She is also a minimally marked sabino, as evidenced by the pattern of her white markings.

Chapter I

An Introduction to Horses

Before we begin to study the vast assort-ment of horse breeds, colors, markings, and coat patterns, we should first explore the basics of the horse itself. In this chapter, we will discover the differences between horse types, discuss the basics of good conformation, and learn about equine body parts, gaits, and registration.

Basic Differences in Type

A horse is a horse (of course, of course!) but there are subtle variations in their basic type. Light horses are by far the most common; these horses typically range in size from 14.2hh up to 17hh, with the

vast majority falling in the 15 to 16hh range. The light horse breeds are ideal for riding and are well suited to a wide range of equestrian disciplines, from trail riding and eventing to dressage and jumping. They are attractive creatures, graceful and elegant, with the resilience and energy to perform in a variety of spheres. Some of the most popular horse breeds are light horses.

Throughout history, draft horses were utilized as work horses. They pulled the plows and assisted with farm chores, steadily performing their daily tasks with strength, stamina, and more than a bit of good sense. Today's draft horses usually don't work quite as hard

Light horses are noted for their riding quality, and light horse breeds rank as some of the most popular in the United States. Some of the most popular breeds include the American Quarter Horse, the Thoroughbred, the Appaloosa, and the American Paint Horse.

Heavy draft horses are an impressive sight! With their massive bodies, heavy muscling, and substantial bone structure, the draft breeds are easily differentiated from the light horse and pony breeds.

as they did in the olden days, but the breeds still possess the inherent qualities and characteristics that make the draft breeds so unique. Massive substance and bone, coupled with immense muscling and strength, are the hallmarks of the draft breeds. These breeds can range in height from as small as 15hh all the way up to over 19hh, with most falling in the 17hh range.

The pony breeds are in a class of their own, immediately distinguishable by their diminutive size (most range from 11hh to 14.2hh), but also by their distinctive pony characteristics. Attractive heads, deep bodies, and a shorter leg-to-body ratio all combine to create a compact package of pony charm. Ponies are well known for their hardiness, strong constitutions, and longevity.

Ponies are immediately recognizable by their deep bodies, shorter legs, and endearing faces. Unlike the light and draft horse breeds, which can range from 14.2hh to over 18hh, pony breeds stand less than 14.2hh and are often much smaller.

The Life Cycle of a Horse

Foal

With its long legs and fuzzy baby coat, a foal is an adorable creature full of spirit and curiosity.

Yearling

Some horses can go through a "gangly" period during their yearling year as they continue growing and maturing. Sometimes the yearling becomes a bit unbalanced in its conformation as it grows, but this resolves itself as the horse continues to mature.

Age Four

By the age of four, most horses are considered to be mature, although some breeds are later to mature than others. A four-year-old should be finished (or nearly finished) growing, but some horses continue to grow until the age of six.

Age Seven

By the age of seven, a horse is considered to be fully mature and in its prime.

Age Twenty

Horses sometimes begin to show grey hair around their eyes as they age (obviously this is not visible on a grey horse!), and sometimes an elderly horse will begin to exhibit signs of a swayback. Their movement can become somewhat stiff if arthritis is present, and it can be difficult for older horses to maintain their weight due to dental problems.

The Essentials of Good Conformation

The basics of good conformation are somewhat universal within all breeds. While the fine points can differ depending on the fundamental characteristics of each breed, the basics of balance and quality transcend to all breeds. Balance is an important element of good conformation. The horse's body should appear equally balanced from head to tail with the neck, midsection, and hindquarters combining together to create a harmonious whole. A horse with a short neck that is accompanied by a long back and weak hindquarters does not create an appearance that is as balanced as it would be if the horse had a longer neck with an adequate length of back and strong hindquarters.

Basic fundamentals of good conformation include large eyes, appropriately sized ears, a properly set neck (not a ewe-neck or bull-neck), and a sloping shoulder. A short back and strong hindquarters are also important.

There is some variation as to the proper set of the tail. Some breeds, such as Arabians, are noted for their high tail-sets and many display an almost level croup. Other breeds, such as the Spanish Colonial horse, are known for their low tail-sets. Most horse breeds possess a balance between the two extremes with tails that are well set without being excessively low or high.

Correct legs are another important aspect of proper conformation. Cow hocks, pigeon toes, and sickle hocks are conformational faults that are commonly seen in a wide variety of breeds.

In addition to proper conformation, the importance of masculinity and femininity should not be underestimated. Male horses should be obviously masculine, while female horses should exhibit the necessary characteristics of femininity. Females are noted for having more refinement through the neck and are more likely to display added length through the back. Male horses are notably cresty through the neck with more substance and muscling overall.

Balance is one of the most important components of good conformation. A horse's front end, midsection, and hindquarters should all be equally balanced without any one portion being excessively longer (or shorter) than the other portions.

Good conformation is essential in a breeding animal, and it is very important for a performance horse. The ability to perform well under saddle or in harness is highly influenced by a horse's conformation.

Making the Grade: Differentiating between Registered Horses and Grade Horses

Generally speaking, the term *grade horse* refers to a horse that is not registered with any breed registry. Sometimes this is because the horse is of unknown parentage, and sometimes it is because the horse is a crossbreed that does not meet the registration requirements for a purebred association. If a horse or pony (let's use the example of a Welsh Pony) has been separated from his registration papers and they cannot be traced, then he would be referred to as a grade Welsh, even though he is purebred.

There are two types of breed registries: those with open stud books and those with closed stud books. Despite the implication, a closed stud book does not mean that the registry is closed to

further entries, but rather that the stud book is closed to any outside influence. In these cases, only the offspring of purebred, registered animals are eligible for registration. On the other hand, an open stud book (frequently seen in newer breeds) allows registration of equines that meet the physical requirements for registration but might not be 100 percent purebred. Some registries will allow horses or ponies of other breeds to be registered as long as they meet the necessary criteria for registration.

Stud books include the registered names and registration numbers of all entries, along with information about each horse's color, markings, date of birth, sire, dam, and owner. These volumes are invaluable for the serious breed enthusiast who can then trace extended pedigrees or track information on color, markings, etc. Some breed registries are moving toward online stud books in lieu of printed and bound stud books.

Registration papers provide a record of important information, including a horse's birth date, registration number, color, markings, and parentage. Stud books contain information on registered animals and are very helpful for tracing pedigrees.

Horse Gaits

Walk
A slow four-beat gait.

Trot
A moderately fast two-beat lateral gait. Some breeds are noted for their high knee action at the trot (the Hackney is a prime example), while others exhibit a longer, lower stride with less knee action.

Canter
A comfortably fast three-beat gait. When circling to the left, the horse should be on the left lead (leading with the front left leg). When circling to the right, the horse should be on the right lead (leading with the front right leg).

Gallop

A very fast four-beat gait.

Other Gaits

Gaited breeds such as the Tennessee Walking Horse, the Rocky Mountain Horse, and the Paso Fino exhibit additional gaits ranging from the Fox Trot to the Paso Largo. The Tennessee Walking Horse's signature gait, the running walk, is displayed in this photo.

Chapter 2

Beyond the Basics

Now that we've explored the basics of horses, it's time to take a moment for a bit of fun before we move on to colors, markings, and breeds. In this chapter we will delve into horse facts and proverbs, see horses in art, and reminisce about some classic children's horse stories. Enjoy!

Horse Facts

- Twins are extremely rare in horses; nearly all mares produce single foals.
- The average normal body temperature of an adult horse is 99.5 to 100.5 degrees Fahrenheit. The average heart rate of a resting horse ranges from 25 to 45 beats per minute.
- The average length of gestation in horses is 340 days, although this can vary by as much as 30 days on either side and still be considered normal. In rare cases, mares have delivered as early as 289 days and as late as 417 days while still producing healthy foals, but of course, these are exceptional cases.
- Some of the smallest horses ever recorded were Miniature Horses that stood less than 18 inches high. One Miniature Horse named Pumpkin stood only 14 inches. On the other hand, the largest horse ever recorded was a Shire that stood 21.2½hh.

- Female foals are known as *fillies*. Male foals are known as *colts*. A female horse over the age of four is known as a *mare*, and a male horse over the age of four is known as a *stallion* (unless he has been castrated, in which case he is called a *gelding*). The proper term for a horse's mother is its *dam*; its father is its *sire*.
- Horses are measured in hands, which is a unit of measurement that equals four inches and is represented as "hh" after the horse's height. Thus, a horse that is 16hh is actually 64 inches tall (horses are measured at the highest point of the withers).

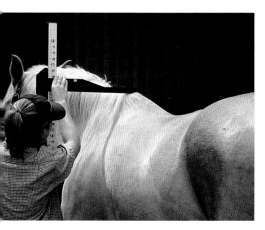

Horses in Children's Literature

For the many horse-crazy children who grow up without the opportunity to own a horse, books are the gateway to experiencing the joy of horses. Who can forget Marguerite Henry's classic tales in her fascinating books *King of the Wind, Justin Morgan Had a Horse*, and *Misty of Chincoteague*? How many starry-eyed girls have dreamed of purchasing a Chincoteague pony on Pony Penning Day, or imagined the excitement depicted in *Born to Trot*?

The perennially popular Black Stallion series by Walter Farley combines the mystique of Arabia with the action-packed drama of horse racing, as readers follow the Black Stallion through a series of books. Also ranking high on the list of classic horse titles is Mary O'Hara's *My Friend Flicka* and its sequels *Thunderhead* and *The Green Grass of Wyoming*. Then there is *National Velvet*, a book made famous by Elizabeth Taylor's performance in the movie of the same name. Other popular classics include Anna Sewell's *Black Beauty* and Will James' *Smoky the Cowhorse*, while less commonly known titles include *Scarlet Royal* by Anne Emery, *Golden Sovereign* and *Dark Sunshine* by Dorothy Lyons, *Spurs for Suzanna* by Betty Cavanna, and *Kentucky Derby Winner* by Isabel McLennan McMeekin. The works of Jean Slaughter Doty, though not as well known as those of Marguerite Henry, are sure to please any young horse enthusiast. Doty's *Summer Pony, Winter Pony*, and *Can I Get There by Candlelight?* are excellent and deserve great praise. Patsey Gray's horse stories are also widely enjoyed, from *Jumping Jack* to *Horsepower*.

For the youngest readers, Walter Farley's classic story, *Little Black, a Pony*, is a delightful tale, accentuated by the pleasing artwork of James Schucker. C. W. Anderson's *Billy and Blaze* is another charming picture book that has been loved by generations of readers.

While today's readers undoubtedly turn to the same classic horse stories that have been enjoyed for decades, they also have the opportunity to peruse newer horse stories. The Phantom Stallion series by Terri Farley is currently very popular, as are the Saddle Club books (and the spin-off television series) and the *Thoroughbred* series.

Children's horse stories are perennially popular with young readers. Classic horse books, such as those written by Marguerite Henry and Walter Farley, are still incredibly popular today.

Horses in Art

Livestock portraiture was a very popular type of art during the nineteenth and twentieth centuries, and paintings of sheep, cattle, and pigs were created in proliferation. Horses were no exception, and portraits of well-known stallions were often painted in order to commemorate these individual horses. As was the case with other types of livestock art, equine artists sometimes embellished their subjects, representing them as even better than they truly were. This embellishment was usually done at the request of horse owners who wanted their animals portrayed in the best possible fashion for promotion and posterity.

The legendary printmakers Currier and Ives did a great deal to popularize art for the American home, and their horse-and-sleigh prints were exceedingly popular. Currier and Ives also produced prints of famous American horses of the era, including a wonderful print of the great stallion Messenger, who was very influential in the foundation of several American breeds, including the American Saddlebred and the American Standardbred.

Equine art is highly collectible today and is a popular way for horse enthusiasts to decorate their homes. While it can be difficult (and expensive!) to obtain original antique art, it is relatively easy and inexpensive to locate reproduction prints of equine art. Also, don't overlook the varied types of equine art, from bronze figurines to hooked rugs.

One of Currier and Ives' popular horse prints, "Imported Messenger," depicts an influential stallion who had an extensive effect on several breeds that were developed in North America.

Horse Proverbs

A few words of wisdom!

"Look not a gift horse in the mouth."

"A nod's as good as a wink to a blind horse."

"Don't change horses in midstream."

"If wishes were horses, beggars would ride."

"It's no use locking the stable door after the horse has bolted."

"Only fools and horses work."

"You can lead a horse to water, but you can't make it drink."

"A horse may stumble though he have four legs."

"For want of a nail the shoe is lost, for want of a shoe the horse is lost, for want of a horse the rider is lost."

"No hoof, no horse."

"When the manger is empty, the horses fight."

"Don't put the cart before the horse."

Chapter 3

Identifying Horse Colors

Judging horse shows is rewarding in many ways, but I have to admit that judging the leadline class is always a highlight for me. Tiny children, bursting with importance in their fancy riding clothes, ride around the ring on a horse led by an adult. When the class is over, I like to take a moment to ask each child a question. Sometimes I ask them their age or the name of their horse, but usually I ask them if they can tell me the color of their horse. The answers never fail to make me smile, as nine times out of ten the child will promptly answer "brown." It doesn't matter if the horse is black, bay, chestnut, dun, or truly brown, the answer is nearly always "brown."

Leadline riders notwithstanding, horses actually sport a broad range of colors and shades—from grey to roan to black to spotted and everything in between. As a horse enthusiast, you will certainly want to be able to promptly and properly identify the colors of horses that you see at shows, on trail rides, and on the Internet. Our equine color guide has been designed to help you do just that. Knowing the names and characteristics of the various shades and colors will take you well on your way toward a solid understanding of equine colors, and also will teach you a bit about color genetics. Brown is just the beginning, so let's explore the equine rainbow!

Colors galore! Unlike some breeds, that are only one particular color, such as the Friesian or the Cleveland Bay, Miniature Horses can be found in a wide array of colors, as seen here.

Equine Color Genetics

Before you cover your eyes and run away screaming at the thought of studying genetics, let me reassure you that the basics of equine color genetics are not at all difficult to comprehend. The knowledge of a few important rules will be of immeasurable help when you're trying to identify breeds or colors of horses.

Equine color genetics is a broad subject with many details that are fascinating to study, but for the purpose of this text, a simple list of basic rules will easily suffice.

A grey horse must have at least one grey parent. This is an important concept. Grey is a dominant gene, and therefore the grey modifying gene cannot skip generations. It's a common misconception that a horse with many grey horses in its extended pedigree will somehow have a higher chance of being grey, but in reality, a horse with two non-grey parents has zero chance of being grey, regardless of the number of greys in its extended pedigree. Roan coloring and the tobiano pattern are also controlled by dominant genes;

One of the basic rules of color genetics is that two chestnut horses will always produce a chestnut foal. Other rules to remember: two black horses cannot produce bay, and a grey horse must have at least one grey parent.

These three young fillies exhibit three different colors. On the left is a black roan (also known as a blue roan), in the middle is a buckskin roan, and on the right is a bay.

therefore roan horses and horses with the tobiano color pattern must have a parent with the same. Roan and tobiano do not skip generations.

Chestnut crossed with chestnut produces chestnut every time. Just as it is with blue eyes in humans, chestnut coloring in horses is controlled by a recessive gene. When both parents possess only the recessive gene, all of their offspring do as well. Two chestnut horses can therefore only produce chestnut foals.

Black crossed with black produces black . . . and chestnut? It seems logical that a black horse bred to another black horse would produce a black foal, and most of the time they do. However, if each black horse is also carrying a recessive chestnut gene, they also have a 25 percent chance of producing a chestnut foal. This comes as a surprise to many horse owners who have been unexpectedly presented with a chestnut foal from their black mare and stallion. Genetic tidbit: Although two black horses can produce chestnut, they cannot produce a bay foal. This is because bay coloring is partially controlled by a dominant agouti extension gene, and black horses only possess recessive agouti genes.

Phenotype is different from genotype. The appearance of a horse is its phenotype, and this is the color that is outwardly manifested. The genotype is the genetic combination that the horse carries, which may or may not be visible in the horse's actual appearance. For instance, consider an example of a grey mare that was born palomino. By the time the mare is five years old, she is a very pale grey. Let's say the mare is bred to a bay stallion and 11 months later, a buckskin foal arrives. If you were judging by phenotype only, you would wonder how a grey horse and a bay horse produced a buckskin. But since you know that the mare was foaled palomino, it is no surprise that she would produce a foal with a dilute gene, since you know her genotype as well as her phenotype.

Equine Color Guide

William Shakespeare, in his play *Henry V*, described a horse this way: "He's of the colour of the nutmeg." Poetic though this may be, we believe that you are probably seeking some color information in this book that will be a bit more specific than that! There are many books that are dedicated solely to the subject of equine colors and genetics; therefore, it was challenging to consolidate such an expansive topic into a single chapter within this book. The task of determining the list of colors to include in this guide was equally challenging. It was immediately apparent that we would be unable to thoroughly cover every color variety and shade, and thus we were forced to make some decisions in order to provide a clear and concise listing of the colors that are frequently encountered in horses. Trying to narrow our list to fit our space limitations was a difficult and time-consuming process, and we hope we have not offended any fanciers of the more unusual colors that we were unable to fit into our listing. We have, however, attempted to include as many colors as possible, and each color listing includes the name of a color and a basic description of the color, along with a photo that illustrates the color. Pertinent information on the genetic factors influencing color is also included. A world of colors awaits, so let's begin!

A Foal of a Different Color

Identifying the color of an adult horse is relatively easy in most cases. It's quite a bit trickier to identify the color of a foal, and in some cases, the foal must be several months (or even over a year old) before their final color is fully apparent.

Foal colors are often very pale in comparison to their mature adult colors. Bay foals can occasionally be mistaken for chestnuts due to the light "baby fuzz" on the foal's legs, which masks the black points. However, close inspection of the foal's mane will typically settle the question. Black hairs at the base of the mane indicate a bay foal. Flaxen or red hairs indicate a chestnut.

The presence of modifying genes, such as grey or roan, is very influential in the appearance of a foal's coat color. Foals that will eventually turn grey are typically born a darker shade than their non-grey counterparts. One rule of thumb is that a bay foal that is born with fawn-colored legs will stay bay; a bay foal that is born with black legs will turn grey, assuming, of course, that at least one parent is grey.

Let's say that you're looking at a jet-black foal. In all likelihood, he's going to be grey because true black foals are usually born a mousey-grey color that sheds out to black. Let's say that you're also looking at the foal's parents and neither one is a grey. This solves part of the mystery immediately, because if neither parent is grey, the foal cannot be grey. The next question would be whether or not either parent is a roan. Foals typically do not exhibit the roan pattern until they are several weeks old and are born a very dark version of their base color, which in this case is black. If neither parent is roan, the foal cannot be a roan. If both modifying genes are ruled out, the foal will stay black. An interesting side note that I have observed in the Welsh breeds is that black Welsh Mountain Ponies are generally born the traditional shade of mousey grey, while black Welsh Cobs are often born jet-black rather than the mousey grey.

This chestnut foal is going to be grey upon maturity. Both of this foal's parents are grey, and you can see the white hairs showing around the colt's eyes that indicate the presence of the grey modifying gene.

This chestnut foal is going to stay chestnut. Both of his parents are chestnut, and there are no white hairs scattered around his eyes.

Seasonal Effects

This yearling colt is sporting a very thick winter coat, which has affected his usually rich coat color and changed it to a rather dull version.

Another aspect of horse color identification is to always keep in mind the seasonal changes that affect the appearance of a horse's coat color. A long, wooly winter coat is very different from a short, shiny summer coat. For this reason, a horse's color can change drastically throughout the year. This is particularly true of certain colors, such as palomino, buckskin, and the roan colors. A palomino in the early spring can be a dark gold that lightens throughout the summer and autumn months before the pale, creamy winter coat comes in. Likewise, a roan can go through several changes throughout the year, particularly when its winter coat is coming out or growing in. A black roan can change from nearly white with black points to a solid steel grey to a nearly purple color during the seasonal coat changes. Generally speaking, winter coats are pale and dull in comparison to the rich, vibrant summer coats.

The same colt is in his sleek and shiny summer coat. Note that the color of this summer coat is deeper and richer.

Bay

Bay, one of the fundamental equine colors, is found in the vast majority of horse and pony breeds. Bay is characterized by a reddish-brown body color accompanied by black legs, ears, mane, and tail. There are many shades and varieties of bay, ranging from a rich mahogany bay to blood bay to cherry bay. Wild bay is characterized by black leg markings that extend only partially up the leg. Bay crossed with bay can produce bay, black, chestnut, or brown, depending on the particular genes of each parent.

Blood bay

Mahogany bay

Red bay

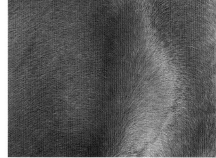

Close-up of red bay

Black

Black, one of the most easily recognizable equine colors, is exactly as its name implies: Solid black all over the body, including the legs, mane, and tail. It can, however, be found in two main varieties: fading black and non-fading black. The basic difference is that a fading black will lighten to a dusky brown shade during the summer months

when the sun is particularly intense. Non-fading blacks remain jet-black in color despite exposure to the sun.

Brown

Brown, sometimes known as "seal brown" within certain registries, can be easily confused with bays and blacks. Brown is, quite simply, a dark brown with black points and a lighter colored muzzle. The genetics behind the inheritance of brown is still not completely understood, but it is an attractive color that is not as common as bay or chestnut.

Buckskin

A cream-colored body with black points (legs, ears, mane, and tail), the buckskin can range widely in shade. Sometimes the body color is a pale cream, occasionally it is more of a dark tan, and at times the body color is a deep brown with golden highlights. Buckskin coloring is occasionally accompanied by amber-colored eyes. Genetically, buckskin is produced by a dilute (cream) gene that lightens a bay horse to buckskin.

Champagne

Champagne is a rare range of color that is often found in the Tennessee Walking Horse breed, as well as in the American Cream Draft. Champagne is a pale golden color that could easily be confused with palomino, dun, or cremello, depending on the shade. Genetic testing has proven that champagne is its own separate color, genetically distinct from the dilute (cream) colors. Champagnes possess amber-colored eyes and pinkish-purplish skin that is sometimes called pumpkin skin. It is the presence of these vital characteristics that can help identify and distinguish champagne from other similar colors.

Chestnut

Chestnut is a very common color that is uniform orange-red over the body and legs. The mane and tail can vary in shade from dark red to nearly white. Chestnuts with red manes and tails are sometimes referred to as sorrel, while some breed registries utilize the term *sorrel* in reference to any chestnut horse. A pale, nearly white mane is called flaxen, and chestnuts with this mane coloring are sometimes referred

to as flaxen chestnuts. On the other hand, liver chestnut is a shade of chestnut that is such a dark color that it sometimes appears to be almost black. Chestnut crossed with chestnut results in chestnut 100 percent of the time.

Flaxen chestnut

Red chestnut/sorrel

Liver chestnut

Cremello

Cremello is one of the double dilute colors that is produced when an otherwise chestnut horse receives a copy of the dilute (cream) gene from each parent, which lightens the horse from chestnut to palomino to cremello. Unlike most horse colors, cremellos have pink skin and always have blue eyes. In fact, some registries do not recognize the term *cremello* and instead refer to the color as *blue-eyed cream*, or BEC. Some registries frown upon the color, and many people have mistakenly referred to cremellos as albinos due to their pale appearance.

Dun

At first glance, you might confuse a dun with a buckskin, due to the similarities in their coloring. Compound this by the fact that some breed organizations register their animals as dun when they are actually buckskin, and you can see how the two colors can be easily confused. An important char-

acteristic to remember is that horses in the dun family, regardless of whether they are apricot dun, clayback dun, red dun, or zebra dun, should exhibit some type of primitive markings, typically a dorsal stripe, but also leg barring. The dun colors are controlled by an entirely different gene than buckskin. The dun gene controls the former and the dilute gene controls the latter.

Grey

Grey is technically known as a modifier rather than an official color, because the grey coloring modifies the base coat of the horse. Despite a great deal of confusion to the contrary, grey horses can be born any color, although many people mistakenly believe that all grey horses are born black. Turning grey is a gradual process, and it can take many years before a grey horse changes to completely white. There are many different terms for shades of grey, including dapple grey (a grey coat covered with dapples), flea-bitten grey (grey with speckles of darker color throughout the coat), and rose grey (commonly used for chestnut horses that are turning grey; the combination of orange/red hairs with white gives the illusion of a rose-grey color). Greys can be either homozygous grey (two copies of the grey gene) or heterozygous grey (one copy of the grey gene). A homozygous grey horse must have two grey parents in order to have inherited two copies of the grey gene, and a homozygous grey horse can only produce grey foals. A heterozygous grey horse could have one grey parent or both grey parents, but because it only inherited one copy of the grey gene, only 50 percent of its foals will be grey.

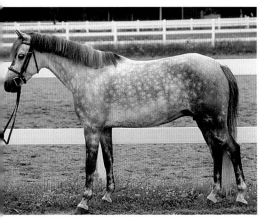

Dapple grey

Aged grey horse that has turned white

Flea-bitten grey

Grulla (also known as Grullo)

Grulla, pronounced "groo-ya" or "groo-yo," is an extremely popular but very rare color that is related to the dun family. It is technically a combination of the dun gene on an otherwise black horse. This results in an eye-catching color combination of black points accompanied by a mousey-grey body color, along with the primitive markings for which duns are well known. The variation in the name of the color stems from the name's Spanish origin. The term *grulla* would technically be used to describe a female horse with the coat color, while *grullo* would refer to a stallion or gelding of that coloring.

Palomino

Palomino is ideally the shade of a newly minted gold coin but it can actually vary in color from an extremely pale cream to a rich orange. Their golden coats are accentuated by a white mane and tail, although intermixed flaxen and black hairs are common in a palomino's mane and tail. Palomino is produced by the action of the dilute (cream) gene on a chestnut base. Palomino bred to another palomino produces an average of 50 percent palomino, 25 percent chestnut, and 25 percent cremello.

Perlino

At first glance it's easy to mistake a perlino for a cremello, as they certainly share many similar characteristics in terms of their appearance, including pink skin, blue eyes, and a pale cream coat. Upon closer inspection, you will undoubtedly notice that the perlino possesses a darker, reddish tint to the mane, tail, and points. This is because the color is a double-dilute on bay. Unlike cremello, which is double-dilute on chestnut, the presence of the darker points on the bay base coat is still slightly visible on the perlino.

Roan

Roan is a modifying gene, which alters the base coat color to include scattered white hairs throughout. The head and legs are typically unaffected by the action of the roan gene. Roans can vary widely in appearance during the year, depending on the season. Roan is a dominant gene that can affect any base color, which means that any roan horse must have at least one roan parent. Roan in the homozygous form is believed to be lethal, as there have been no documented cases of homozygous roan horses. Genetically speaking, it is proper to refer to roans in conjunction with their base color (bay roan, chestnut roan, palomino roan), although there are many traditional terms that are still used today (blue roan, strawberry roan, red roan) that you will undoubtedly encounter from time to time.

Silver Dapple

Widely known for its frequency in the Shetland pony breed, silver dapple is found in several other breeds as well. The *dapple* part of the name can be a bit confusing, as dappling is not necessarily present in all instances. Silver coloring is often confused with bay, but an abundance of flaxen hairs in the mane and tail is a good clue to the presence of the silver gene. Brown legs accompanied by a bay body color is another indicator of the silver gene.

Smoky Black

We've already learned that palomino is produced by the dilute gene on a chestnut horse and that buckskin is produced by the dilute gene on a bay horse. Smoky black is the result of the dilute gene on a black horse. Smoky black is easily overlooked, and many smoky black horses are incorrectly labeled as black, simply because the presence of the dilute gene does not dramatically affect the black coloring. Sometimes the only clues are cream-colored hair in the ears and amber-colored eyes, with the horse appearing entirely black otherwise. However, if a black horse has a parent with a dilute gene, it's always a good idea to examine the horse's ear hair. If you find golden-colored hair, it's extremely possible that the horse inherited a dilute gene and is truly a smoky black.

Chapter 4

White Markings and Other Identifying Characteristics

Sometimes white markings do not meet the criteria for the traditional "star, stripe, snip, or blaze" definitions. This Miniature Horse exhibits an irregular star that continues down his face, veering sharply down the left side of his face.

By now you are well equipped with the information necessary to identify a horse's color. One glance at a reddish-brown horse with a black mane and tail and you immediately recognize that he is bay, and by now you have even acquainted yourself with fancy terms like *grulla* and *dilute gene*. Now it's time to discover several

other distinguishing characteristics that can assist you with horse identification. Eye color, hoof color, white markings, whorls, and brands are various aspects that combine to make each horse a unique individual. A brief overview of the general terminology associated with these factors will help to demystify any confusion over the differences between coronets and stockings, amber eyes and brown, black hooves or white, and an assortment of other features.

Eye Color

The majority of horses have dark-brown eyes, and it is the eye color that you will encounter most of the time. There are, however, circumstances in which the eyes can be another color. Amber eyes, a golden color, are often found on horses that possess a copy of the dilute (cream) gene, such as palomino, buckskin, or smoky black. Horses with champagne coloring also exhibit amber eyes.

Blue eyes appear in horses with two copies of the dilute (cream) gene, such as cremellos, perlinos, and smoky creams. Horses with extensive white facial markings are also noted for having blue eyes if the white marking extends onto or over the eye. Occasionally a brown eye that has a splash of blue in it will be found; this is usually the result of a facial marking touching the eye in only one location. Note: It's important to realize that foals, like human babies, are often born with deep blue eyes.

This is not necessarily indicative of their adult eye color, as many foals' blue eyes darken to brown after a few months.

This horse's eye is blue as a result of a white blaze that extended over his eye. The blaze is no longer visible because the horse has gone grey.

A brown eye, which is the most common eye color in horses.

At first glance you might confuse this eye color with blue, but this is actually a "wall-eye," an eye color that is characterized by white coloring around the pupil.

Whorls

The *Encarta World English Dictionary* defines a whorl as "something spiral-shaped: something in the shape of a spiral, coil, or curl." Whorls in horses are patterns of hair that are arranged in a spiral shape. Quite often they are located at the center of the forehead, but many horses also have whorls in other locations. The presence of whorls can provide another aspect of identification. Folklore attributes a great deal of importance to whorls, including their placement, the direction of the spiral, and the number of whorls a horse possesses. For instance, some people say that a forehead whorl is indicative of the horse's disposition. Highly placed whorls are supposed to signify a spirited personality, and whorls that are placed lower on the forehead indicate a calm, quiet demeanor. Others claim that the direction of the spiral (clockwise or counter clockwise) can indicate whether a horse is right-sided or left-sided (similar to right- and left-handedness in humans). While this is obviously anecdotal data, the fact remains that whorls are a unique part of each horse's individual appearance, and it is useful to make note of them.

A whorl is a circular swirl of hair. Whorls can be located anywhere on the body, but are commonly seen on the forehead.

Brands

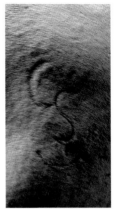

This brand is a good example of the type of brand often used by private farms to mark their stock. If a horse were to become separated from its registration papers, a brand such as this one could help trace a horse's identity.

A freeze brand, as seen on the back of an Irish Draught horse.

Some breeds are branded by their breed registry with an identifying mark on the body, distinguishing them as belonging to that particular breed. If you are familiar with the appearance of the brands that are utilized by the various registries, you will immediately be able to identify a horse's breed simply by checking the brand. Brands are commonly used by the Warmblood breeds, but other breed organizations also use them, including the North American Sportpony, the Shagya Arabian, and the Haflinger.

While Thoroughbreds are not branded, their Jockey Club registration number is tattooed on their upper lip, which not only identifies them as a Thoroughbred but also gives an option for researching the horse's name, bloodlines, date of birth, etc.

Additionally, some horse owners have their horses branded or tattooed in an effort to make them easily identifiable if they become lost or stolen. Sometimes breeders that produce large numbers of foals each year also brand their horses in order to signify that they come from that particular farm.

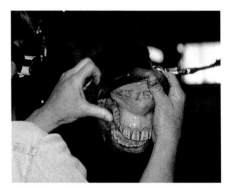

A veterinarian is checking the lip tattoo on this Jockey Club–registered Thoroughbred. Permanent identification with a breed organization can help to track a horse's age and pedigree.

This is another type of identifying brand, pictured here on the hip of a newly branded Latvian foal. The brand signifies the foal's registration with the Rheinland Pfalz-Saar registry.

Hoof Color

A close-up of a black hoof. Historically, black hooves were valued for their strength and durability.

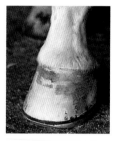

A white leg accompanied by a white hoof. People once believed that white feet were weaker than black feet, but this theory has been disproved.

Dark ermine spots along the coronet band are the cause of the dark stripes in this otherwise-white hoof. Striped hooves are commonly found in Appaloosas, but also in many other breeds.

Folklore has long held that white hooves on horses are somehow weaker than black hooves, and those with white markings (and thus, white hooves) were often discriminated against in the past. This theory is no longer considered accurate, as studies have shown that black hooves are no stronger than white hooves. Today's experts believe that hoof color does not play a part in the quality or strength of the hoof itself and is only indicative of whether or not the leg has a white marking.

Along with black (sometimes called blue) and white hooves, there are also striped hooves. While striped hooves are a major characteristic in the Appaloosa breed, they are also found in a multitude of other breeds. A striped hoof often occurs because of ermine spots on the coronet band.

Feathering

Heavy feathering is a characteristic of many horse and pony breeds, including the Shire, Clydesdale, Friesian, Gypsy Vanner, Welsh Cob, and Fell Pony (shown here).

Hoof color is generally related to leg color, meaning that a leg with a white sock or stocking usually is accompanied by a white hoof. Legs that are devoid of white markings are usually accompanied by a black hoof. The exception is when dark ermine spots are present around the coronet band adjacent to the hoof. In these unusual cases, there can be a white stocking present, but the ermine spots alter the color of the hoof so it becomes black instead of white.

Feathering is commonly seen in native pony breeds and is also found in many draft horse breeds and some light horse breeds (such as the Friesian). Feathers are long hair that grows on the lower legs and fetlocks. While most horses have a bit of hair on their fetlocks, it is usually clipped off. Breeds that are noted for abundant feathering, such as Shire horses or Dales ponies, do not traditionally have their feathers clipped and instead exhibit in a natural state at horse shows.

White Markings

An extensive white face marking sometimes extends up underneath the horse's chin, such as on this chestnut horse.

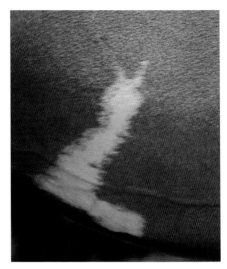

Horses with extensive white markings on their face and legs often exhibit belly spots. Small patches of white markings often occur under the belly and occasionally on the horse's side. Some belly spots are only visible when the horse is rolling.

One of the most distinctive characteristics of any horse is its unique combination of white face and leg markings Many horses also exhibit white marking patterns on their body (see Chapter 5 for a complete discussion of pinto and Appaloosa coat patterns), but for now let's concentrate on the various markings that are commonly found on the face and legs.

There is an old poem that offers a bit of advice regarding white markings and horses: "One white sock buy him / Two white socks try him / Three white socks think for a day / Four white socks turn him away."

This poem harkens back to the days when white socks reflected negatively on a horse's quality. Today, white markings are more popular than ever with many horse enthusiasts preferring the added bonus of white markings on their horses.

The exceptions to this are breeds such as the Friesian horse and Exmoor pony, whose breed standards prohibit the presence of white markings, with the possible exception of a small star.

The presence of white markings is somewhat controlled by the body color of the horse. Studies have found that chestnut horses typically exhibit more extensive white markings than their bay or black counterparts, with homozygous black horses exhibiting the fewest white markings. Studies have also shown that the amount of facial white correlates somewhat with the amount of leg white. For instance, horses with wide blazes generally have more white on their legs than horses with only a small star.

Let's explore some of the face and leg markings you will commonly see in horses and ponies.

Face Markings

Bald Face

This is a very extensive type of facial marking and involves the majority of the horse's face. A bald face usually encompasses one or both of the eyes and extends down the face, often covering both nostrils and the entire muzzle. A bald face is very distinctive and is often accompanied by blue eyes and extensive leg markings.

Blaze

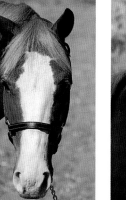

A blaze, a scaled-down version of a bald face, is narrower and usually does not touch either eye. The nostrils may or may not be included in the white area, and the marking does not usually extend down the sides of the face.

Stripe

A stripe is even narrower than a blaze, is very thin, and usually extends from the forehead to the muzzle. It typically does not include the nostrils.

Star

Stars can vary widely in size and shape, but generally speaking, they are a patch of white between the eyes. Some stars are oval, circular, or diamond-shaped, and some are irregularly shaped.

Snip

A snip is a disconnected white patch between the nostrils, usually accompanied by a star. A snip occasionally occurs without a star, but this is unusual.

Sometimes a horse will exhibit more than one type of white facial marking. This horse possesses a star and a snip.

Leg Markings

Coronet

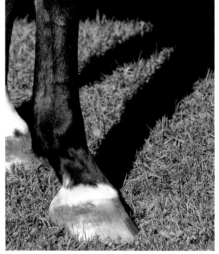

A coronet is the smallest of white leg markings and encircles the hoof's coronet band only.

Pastern

A pastern is a white marking that extends up onto the pastern but does not include the fetlock joint.

Ankle (or Sock)

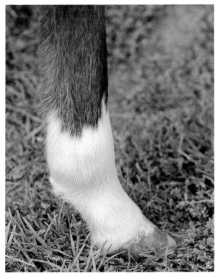

The ankle marking includes the fetlock joint and is the marking typically referred to as a sock.

Half-Stocking

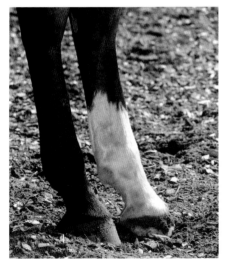

A half-stocking extends past the fetlock and onto the cannon bone but stops about halfway to the knee or hock.

Full Stocking

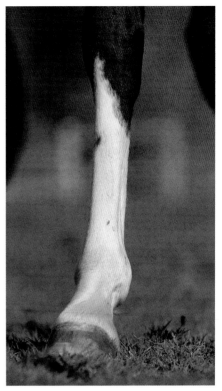

A full stocking is a tall white marking that extends all the way up the cannon bone, reaching or including the knee or hock.

What Do You Call That?

While we have covered the most frequently seen variations in coat colors, coat patterns, leg and face markings, and hoof colors, there are also a few unusual equine characteristics that you may encounter from time to time.

Bend Or Spots

Named for a Thoroughbred stallion who exhibited this type of marking, Bend Or spots (also known as Ben d'Or spots) are small areas of dark hair that appear on chestnut (and chestnut-based) horses.

Brindle

An unusual and very rare type of coat pattern, the term *brindle* refers to dark vertical stripes that are reminiscent of zebra stripes.

Sclera

White sclera that shows along the edge of the eye is one of the distinguishing features in Appaloosas and is often found in horses of other breeds. Some horsemen and women refer to this characteristic as a human eye because the appearance is closer to that of the human eye than the typical horse eye.

Primitive Markings

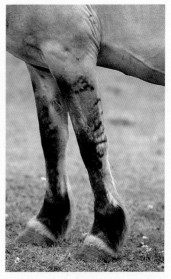

These are found in several breeds and in several colors and consist of a dorsal stripe (also known as an eel stripe) down the back and leg barring (also known as zebra markings).

Prophet's Thumb

Occasionally found in horses of any breed, the prophet's thumb is a small indentation typically located along the side of a horse's neck.

Chapter 5

Coat Patterns

Now that we've thoroughly discussed the basics of equine colors and have familiarized ourselves with the various types of white markings, it's time to branch out into horses that exhibit specific coat patterns.

Undoubtedly you've seen eye-catching horses that exhibit one of the interesting coat patterns that we'll discuss in this chapter. But perhaps you've heard the different terms such as *overo*, *spotted blanket*, or *piebald*, and have wondered exactly what they mean. Or perhaps you've admired an American Paint Horse at a show, wondered which coat pattern it exhibits, and aren't sure what the differences are between a sabino and a tobiano, or a tovero and an overo.

In this chapter, we'll attempt to demystify the various coat patterns. We'll explain the characteristics that differentiate the various patterns found in American Paint Horses, Appaloosas, and other breeds so that the next time you see one, you'll be able to quickly reference this guide to discover the proper name for the pattern.

Paint and Pinto Patterns

The catch-all term for horses that exhibit extensive white markings on their bodies is *pinto*. While this is a generic term for the pattern of horses that display two colors, the Pinto Horse Association does register horses of many breeds that exhibit the appropriate types of white markings. The American Paint Horse is undoubtedly the breed most widely recognized for its distinctive patterns of white color interspersed with their base body color,

but these patterns of white are also found in several additional breeds, from Gypsy Vanners and Miniature Horses to Shetland Ponies and Missouri Fox Trotters.

Historically, horses with pinto coats were referred to in a simple fashion by the use of two terms: *piebald* and *skewbald*. These terms are now considered obsolete and are virtually unused in today's horse terminology. The term *piebald* was traditionally used to describe a horse that

Just like a calico cat, this pinto horse exhibits a mane that is comprised of three different colors. The white hairs (caused by the white markings along the horse's neck) are intermixed with black and flaxen hairs in an unusual way.

exhibited a black coat mixed with white, while *skewbald* was the term used to describe horses that were any other color and white. Today these terms are simply not specific enough to differentiate between the various patterns that exist in colored horses, nor do they keep up with all of the current research and knowledge that exists about each of the different coat patterns.

New research and study is changing the way we look at the various patterns. The sabino pattern, for instance, was scarcely mentioned until recent years, yet now there is a far greater understanding of the characteristics that make up the sabino pattern, and a DNA test has even been developed to check for the sabino pattern.

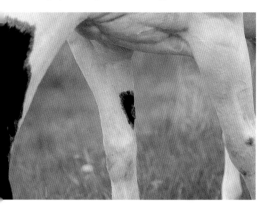

This American Paint Horse mare exhibits a unique dark spot on the inside of her white leg. It's interesting to note that the spot coincides with the exact area of her chestnut (a hard, callus-like spot on the inside of a horse's leg, above the knee).

Tobiano

Tobiano is one of the easiest patterns to recognize once you have familiarized yourself with the characteristics of this coat pattern. One of the most typical characteristics that tobianos exhibit is a dark head with only a small star or narrow stripe. The extensively marked bald faces that are found on some pinto horses are usually only seen on the

Historically, horses with pinto patterns were referred to as *piebald* (black and white) and *skewbald* (any other color and white, as seen here). These terms are now considered obsolete and have been replaced by several more specific terms.

One of the main characteristics of tobianos is the fact that their white markings often extend over their topline, as seen on this Miniature Horse. The face marking on this Mini is not typical of the tobiano pattern, which could signify the presence of the overo color pattern.

overo-related patterns. Even though the traditional understanding of white markings suggests that extensive face markings

A classic example of the tobiano pattern is shown on this Pintabian. If you look closely, you'll see the shaded area where his white markings meet his body color. This horse also exhibits the dark face and white legs that are commonly seen on tobianos.

and high white stockings are usually correlated (with minimally marked faces most common on horses without leg markings), the tobiano pattern seems to completely disregard this rule of thumb. Tobianos, with their conservatively marked faces, often have four white legs, and the white usually extends onto the body.

Another distinctive trait in tobianos is that their white body markings usually cross the topline in at least one location. The white markings may cross over the spine or over the mane, which results in a particularly unique two-colored mane. This is especially noteworthy from an identification perspective, as the white markings on horses with any of the overo patterns (with the exception of the most maximally expressed examples) typically will not cross their toplines.

Tobianos also feature additional characteristics that are quite distinctive.

Genetic testing can determine whether or not a horse is homozygous or heterozygous for the tobiano pattern. Heterozygous tobianos possess only one copy of the pattern, while homozygous tobianos possess two copies. Homozygous tobianos will always produce tobiano offspring.

On some tobianos, when the white markings meet the colored areas, the two colors overlap to create an area of lighter shading. Tobianos also often exhibit ink spots, which are colored spots that appear within the confines of the white patches on the horse's body. There is some research suggesting that the ink spots are mainly found in horses that are homozygous for the tobiano gene.

This tobiano features a two-toned mane, which is caused by the fact that the white markings on his body cross over his neck, thus allowing for a portion of his mane to be white. He features the characteristic white legs and dark face that typify the tobiano pattern.

The tobiano pattern can be minimally expressed, as shown on this Shetland pony gelding. The white spot on his hindquarters and his white tail are the only clues that he carries the tobiano pattern.

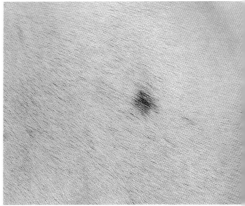

Tobianos often exhibit ink spots, which are tiny dark spots interspersed throughout the white markings in their coats.

Overo Types

In the past, many varieties of overo have typically been lumped together under the overo heading (essentially meaning "not tobiano"), but there are actually several sub-type patterns. Because each one is distinctive and unique, we will present each one separately here for reference.

Frame Overo

At first glance, you might have difficulty in differentiating between tobianos and frame overos, but they are actually quite different when you carefully compare their characteristics. Frame overos, being a part of the overo family, usually exhibit considerable white on their faces that extends over the eyes, onto the jaws, and under the chin. In addition, the markings on frame overos typically do not cross the spine or neck. As the name implies, the white markings appear to be framed on the body within the parameters of the darker coloring.

Unlike tobianos, frame overos often exhibit dark legs. This mare exhibits a uniquely shaped pattern of white markings on her side and neck. Note the shape of the state of Illinois that sits in the center of her white body patch!

Unlike tobianos, sabinos, and splashed whites, all of which regularly feature four white legs, dark legs are commonly found on frame overos.

This gorgeous horse is a wonderful example of the frame overo pattern. As you can see, the white markings are framed by the darker color all along the horse's topline, with the exception of a small place along his mane.

Sabino

Quite distinctive in appearance, the sabino differs considerably from any of the other coat patterns. Sabinos are characterized by their specks and flecks. In fact, the name *sabino* comes from the Spanish word meaning speckled. The sabino pattern can be extensively or minimally expressed. In its most minimal form, a sabino horse might exhibit only a tall stocking and a narrow blaze, while in its most extensive form, a sabino horse might be nearly or completely white.

Minimal to moderate sabinos are characterized by high stockings with jagged or irregular edges, wide blazes that may extend over the eyes (in which case the eyes are usually blue), chin or lip spots, roaning or flecking of white hairs along the flanks, and belly spots or patches. In addition to all of these characteristics, sabinos that are maximally expressed also exhibit extensive body white, along with the basic characteristics of flecking, roaning, speckling, jagged edges, etc.

There is still a great deal that is unknown about the inheritance of sabino. A new DNA test was recently developed that can ascertain the presence of the sabino gene and can also detect whether or not the horse possesses the sabino gene in heterozygous or homozygous form. However, it is still not clear exactly how the trait is inherited, as there are documented cases of minimally marked parents producing a full-blown maximum sabino foal with large amounts of body white.

At first glance, you might not notice this mare's sabino characteristics, but upon closer inspection it's easy to spot the jagged edges of her stockings, particularly her right hind leg (note the spear-like marking). Another obvious sabino characteristic is the white spot on her lower lip and chin.

Identifying a horse's pinto pattern can be tricky, but thankfully there are certain clues to go by. This horse's legs are white, which eliminates the frame overo pattern. She has irregular markings, which eliminates the splashed white pattern. Her face is white, which eliminates the tobiano pattern. This only leaves sabino or tovero, but the horse's speckled and flecked appearance is much more consistent with the sabino pattern than the tovero.

Splashed White

Unlike the white markings of the sabino with its speckled and jagged edges, the splashed white pattern is a pattern made up of clearly defined lines. The dividing lines between the colored areas of the body and the white markings are crisp and clear. In the splashed white pattern, the white areas are typically heavier on the lower portions of the body and usually include all four legs with the white extending up onto the body. A bald face is also commonly seen with this pattern, as are blue eyes.

This pattern is aptly named, as it has often been said that a splashed white horse looks as though it has been splashed with white paint, or more precisely, dipped into a can of paint, with the only colored portions being the topline and the ears. It is a distinctly beautiful pattern and quite unlike any other.

This horse aptly illustrates the splashed white pattern of markings. It looks as if he has just walked through a pool of white paint. He has high white stockings that extend up toward his body and a bottom-heavy white blaze.

Combination Patterns

Tovero

What do you get when you cross a tobiano horse with an overo horse? The tovero (pronounced toe-VAIR-o) is produced when a horse inherits both the tobiano and the overo color patterns. This can be quite complicated to understand when you have just learned the various rules for recognizing tobianos and overos. For instance, you may see a horse with vertically placed white markings, ink spots, and a completely bald face, and begin scratching your head. After all, if the horse displays ink spots he's a tobiano, but if he has a bald face, then he's one of the overo patterns. So what is he? In this case, he is likely a tovero.

There are other terms that you may hear with regard to combination patterns, such as *sabero* (which stands for sabino/overo), but these terms are not recognized by the American Paint Horse Association. This is understandable when you consider that the sabino is actually a sub-type of the overo pattern; therefore, to call it a sabino/overo combination is not really necessary. This is not to say that a horse cannot exhibit a sabino pattern and a splashed white pattern, but it's just not necessary to call it a *splashino* or some other invented term.

A highly unusual example, this tovero mare is so extensively marked that her ears are white. The majority of pinto horses do not have white ears because their ears usually retain the base body color. It has been noted that tovero horses have a tendency toward being more extensively marked than horses with only one pinto pattern. This is certainly the case with this mare.

Appaloosa Patterns

Although they vary in appearance from the pinto patterns that we've already discussed, Appaloosas are also characterized by patterns of white hair intermixed with the horse's base body color. Appaloosas come in a variety of different patterns, but there are several characteristics that apply to each type of Appaloosa. These include mottled skin, prominent white sclera of the eye, and striped hooves.

The patterns of Appaloosa coloring are also found in other breeds, including the Pony of the Americas (which descends from the Appaloosa), the Colorado Ranger, and the Knabstrupper, which originated on a different continent yet shares many of the same color patterns. For the sake of simplicity, the colors described below will be referred to by the terms for which they are known in the Appaloosa breed.

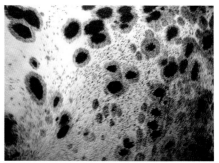

Appaloosas can be found in a wide variety of colors and patterns, many of which involve spotted coats. The Dalmatian-like effect shown here is very striking, with the dark spots appearing prominently throughout the white coat.

This close-up of an Appaloosa coat illustrates the shading that sometimes occurs where the dark spots meet the white markings, resulting in a halo effect around each spot.

In addition to their unique coat patterns, Appaloosas exhibit several very distinctive characteristics, such as the skin mottling and eye sclera illustrated on this horse.

This Appaloosa is a prime example of the wide variety of features—spots, specks, roaning, and mottling—that distinguish the Appaloosa (and the Pony of the America, and the Colorado Ranger) from other horse breeds.

Aptly illustrating the fact that an Appaloosa's base coat can be any color, this horse is a prime example of a roan horse with Appaloosa characteristics (including mottling and sclera).

The Leopard Complex

Aptly named, the leopard complex is just that: complex! All of the various Appaloosa patterns are related and the group of patterns is known as the leopard complex. The most distinctive of these patterns is the leopard pattern. It is easy to remember and easy to identify. The leopard pattern is essentially the Dalmatian of the Appaloosa family, as it consists of dark spots that extend over the entire body and are evenly distributed without particular concentration to one area or another. The leopard pattern is quite attractive with the multitude of dark spots throughout the body.

Another common Appaloosa pattern is the blanket pattern, which can be found in either a solid blanket or a spotted blanket. The blanket is located on the horse's hindquarters and may be completely white (known as the solid blanket) or speckled with dark spots (known as the spotted blanket).

The varnish pattern is somewhat similar in appearance to the traditional roan color found in many breeds of horses. The similarity is due to the fact that both the roan color and the varnish pattern are comprised of white hairs intermixed with the darker base color of the horse. In the varnish pattern, however, there are fewer white hairs along certain portions of the body such as the shoulders, the ears, or the legs, but the face is usually very roaned.

The snowflake pattern is appropriately named because at first glance it truly looks as though snowflakes randomly landed on various portions of its body. These "snowflakes" are small white areas that are found throughout the horse's body.

The leopard Appaloosa pattern is characterized by a white base color accompanied by dark spots distributed throughout the body. This is in contrast to the spotted blanket Appaloosa pattern, in which the dark spots are restricted to the horse's hindquarters.

The spotted blanket pattern consists of a white blanket on the horse's hindquarters speckled with dark spots. A solid blanket pattern is very similar to a spotted blanket pattern, but without the spots.

It is important to remember that any Appaloosa horse may exhibit more than one of these patterns, so identification can be tricky when you are confronted with an example that displays the characteristics of more than one pattern. Of course, that adds to the fun and enjoyment of the patterns!

The varnish pattern is another beautiful Appaloosa pattern. The dark ears, legs, and shoulders illustrated on this horse are common characteristics of the varnish pattern.

Chapter 6

Breed Profiles

How to Use the Breed Profiles

We have arrived at the part of the book that everyone has been waiting for—in-depth profiles of horse and pony breeds. Let's take a moment to explore the various categories that exist within each breed profile so that you can quickly delve into the information about each breed.

Breed Name: This is self-explanatory—the breed name category lists the name of each breed. In some cases, the official breed name differs slightly from the breed name that is sometimes used in conversation. An example of this is the Tennessee Walking Horse, which is the breed's official name as recognized by the Tennessee Walking Horse Breeders and Exhibitors Association. However, Tennessee Walking Horses are often referred to as Tennessee Walkers. In other cases, the breed has undergone a name change (the North American Sportpony was formerly known as the American Sport Pony), and in these cases, the most recent official name is listed.

Breed Description: This is a brief overview of the history of the breed: how it was developed, where it originated, the breeds that were instrumental in its development, and so on. Important physical characteristics of type, color, and markings are also noted.

> **Colors:** Many breed organizations provide color specifications and/or restrictions for their breed, and this category provides information on the appropriate and accepted colors for each breed. Some breeds are found in a veritable rainbow of colors, while others are found in only a single color. For identification purposes, you can cross-reference the breed profiles with Chapter 3.
>
> **Height:** The average height or range of heights typical for the breed. Some breed organizations and registries impose specific height restrictions, while other breed associations do not present any parameters for height.

Opposite: The Icelandic is an ancient breed noted for its hardiness and its longevity. In this chapter we will fully explore a multitude of light horse, draft horse, and pony breeds.

Type: As discussed in Chapter 2, there are three generally accepted horse types: light horses, draft horses, and ponies. Each breed is marked with its particular type.

Gaited: Some breeds exhibit additional gaits beyond (or in replacement of) the typical walk, trot, and canter. We have therefore highlighted the breeds that are gaited.

ALBC Status: The American Livestock Breed Conservancy (ALBC) is a nonprofit organization founded in 1977 to help preserve and promote rare and endangered livestock breeds. The ALBC currently highlights several horse and pony breeds that have global populations of fewer than 10,000, and in some cases, fewer than 2,000 individuals. These breeds are on the ALBC's critical, threatened, and watch lists. The ALBC also maintains recovery and study lists of horse breeds that may not meet the criteria for the other lists, but are still possibly endangered.

RBC Status: Rare Breeds Canada (RBC) maintains an equine conservation list as well, and their criteria are based in part upon the number of purebred females that are registered in a given year. Critical means that between 1 and 15 purebred females are registered each year; endangered means between 16 and 50; vulnerable means between 51 and 150; and at risk means between 151 and 500.

RBST Status: The Rare Breeds Survival Trust (RBST) is a conservation group in the United Kingdom. According to their website, "Our purpose is to secure the continued existence and viability of the United Kingdom's native farm animal genetic resources." To this end, they maintain a watch list for rare breeds of British origin, classifying all breeds that have fewer than 3,000 breeding females.

Breed Organization Website: An excellent source for additional information about each breed. These websites can be a wonderful resource for anyone seeking to learn more about a certain breed. Many of these websites feature photographs, show information, breed history, news, and registration information, along with links to breeders and enthusiasts who can help you learn more about their breed. There is truly a wealth of information available at your fingertips by visiting these websites.

Breed Profiles

American Bashkir Curly

Breed Description: The history of Curly horses—a breed with curly coats and curly manes and tails—is somewhat shrouded in mystery. Some sources believe that the foundation stock of the Curly horses are descended from the horses in Russia's Bashkir region, while others believe that they are related to the Lokai horses of Tajikistan. The first Curly horses in America were found in Nevada, just prior to the turn of the twentieth century, although there is some record of Curly horses within Native American herds in the early 1800s. In any case, this unique type of equine coat differs significantly from the typical smooth coat found in the vast majority of horse breeds. The mane and tail are also curly, giving the effect of ringlets, and Curly horses are often "double maned," meaning that the mane falls on both sides of the neck. According to the American Bashkir Curly Registry (formed in 1971), there are currently fewer than 4,000 Curly horses worldwide. An asterisk in front of a Curly horse's name indicates the presence of a visibly curly coat. This is entirely different from the meaning of the asterisk in front of the names of other breeds, such as Welsh Ponies or Arabians. In those cases, the asterisk signifies that the animal was imported.

Colors: All
Height: 14.3 to 15hh
Type: Light horse
Gaited: No
ALBC Status: None
RBC Status: Critical
RBST Status: None
Breed Organization Website: www.abcregistry.org

American Cream Draft

Breed Description: The American Cream Draft possesses all of the characteristics that the draft breeds are noted for, along with the added distinction of its unique signature color. The legend goes that the breed's matriarch, Old Granny, was found at an Iowa horse auction in 1911. Prized for her eye-catching color and draft type, Old Granny produced several foals that also exhibited her cream coloring, and these horses were the foundation of the American Cream Draft breed. The American Cream Draft Registry was established in 1944; however, it was not until the latter part of the twentieth century that the breed's exposure and popularity began to increase. According to the American Cream Draft Horse Association, there are currently fewer than 400 registered American Cream Drafts in existence. An American Cream Draft horse must exhibit three specific characteristics: amber eyes, pink skin, and a cream-colored coat, while also maintaining the heavy draft characteristics that define the breed.

Colors: Cream (genetically champagne)
Height: 15 to 16.3hh
Type: Draft horse
Gaited: No
ALBC Status: Critical
RBC Status: None
RBST Status: None
Breed Organization Website: www.acdha.org

American Paint Horse

Breed Description: The American Paint Horse Association (APHA) was established in 1962 out of a need to offer registration and showing opportunities for so-called crop-out American Quarter Horses, which were animals that displayed white markings in excessive amounts and were ineligible for registration with the American Quarter Horse Association. At that time, many Quarter Horse enthusiasts were disappointed in the fact that they could not register and show the brightly colored horses that they admired and enjoyed. Since 1962, the APHA has blossomed and the breed is now one of the most popular in the United States. Because of the breed's Quarter Horse heritage, the American Paint Horse is a quality stock-type breed that is a favored choice with Western riders. The offspring of Jockey Club–registered Thoroughbreds are also eligible for APHA registration, provided that they exhibit the necessary amount of white markings.

Colors: The APHA recognizes three official color patterns: tobiano, overo, and tovero. These three patterns may be exhibited on any of sisteen recognized body colors including bay roan, bay, black, blue roan, brown, buckskin, chestnut, cremello, dun, grey, grullo, palomino, perlino, red dun, red roan, and sorrel.
Height: 15 to 16hh
Type: Light horse
Gaited: No
ALBC Status: None
RBC Status: None
RBST Status: None
Breed Organization Website: www.apha.com

American Paint Pony

Breed Description: Given the enormous popularity of the American Paint Horse, it is hardly surprising to learn that there is an American Paint Pony registry to allow registration for colorful equines of pony size. Established in 1972, the American Paint Pony registry is open to all breeds and backgrounds, except for equines exhibiting gaited and/or Appaloosa characteristics. In addition, the registry maintains a height limit of 14.2hh, which makes sense in consideration of the fact that it is a Paint Pony registry. American Paint Ponies make eye-catching mounts for young riders, especially those who are fond of American Paint Horses but are seeking a smaller version.

Colors: All colors, accompanied by recognizable white body color or pinto characteristics
Height: 12 to 14.2hh (Ponies smaller than 12hh are eligible for registration in the Pee Wee division.)
Type: Pony
Gaited: No
ALBC Status: None
RBC Status: None
RBST Status: None
Breed Organization Website: www.americanpaintpony.com

American Quarter Horse

Breed Description: The American Quarter Horse carries the distinction of being the most popular breed in America, which is a tremendous testament to the breed's excellent qualities. It seems fitting that the most popular breed in America is one that was developed right on our own shores. The American Quarter Horse has its foundation in the Spanish horses that were crossed with the early Thoroughbred imports from England. The breed earned its

name from its ability and aptitude in racing at the distance of a quarter mile. Even today, despite the fact that the Quarter Horse's primary use is as a stock-type pleasure horse, the breed is still extremely fast at quarter-mile racing. Quarter Horses are valued for their natural ranching ability and cow sense, and they are frequently found competing in such classes as Western Pleasure, Western Horsemanship, Western Riding, and Reining at horse shows. The annual American Quarter Horse World show attracts an impressive number of entries each year, and the American Quarter Horse Association boasts more than 300,000 members.

Colors: Sixteen colors are recognized: brown, black, grey, sorrel, bay, chestnut, dun, red dun, grullo, buckskin, perlino, cremello, bay roan, blue roan, red roan, and palomino
Height: 14.3 to 16hh
Type: Light horse
Gaited: No
ALBC Status: None
RBC Status: None
RBST Status: None
Breed Organization Website: www.aqha.com

American Quarter Pony

Breed Description: Due to the immense popularity of the American Quarter Horse, it is not surprising to discover that there is also a registry for their more petite counterparts: the American Quarter Pony Association. Despite the inference from the name, Quarter Ponies come from a variety of backgrounds and breeds, not merely Quarter Horses. A Western or stock-type build is considered correct for Quarter Pony registration, but the registry prohibits any pony with Appaloosa or Pinto characteristics, as well as gaited ponies. The registry was formed in the 1960s to allow grade ponies and small grade horses the opportunity to exhibit at horse shows that were restricted to registered equines.

Colors: All colors, but Appaloosa, Pinto, Paint, or Albino ponies are ineligible for registration
Height: 46 to 57 inches (the average Quarter Pony is 13.2hh)
Type: Pony
Gaited: No
ALBC Status: None
RBC Status: None
RBST Status: None
Breed Organization Website: www.aqpa.com

American Saddlebred

Breed Description: One of the best known of the gaited breeds, the American Saddlebred is lauded as being "the ultimate show horse" by the American Saddlebred Horse Association (ASHA). Indeed, the breed's charisma and presence make it one of the most popular breeds in the United States. According to the ASHA, the foundation breeds of the Saddlebred were the Narragansett Pacer and the Thoroughbred (the Thoroughbred stallion, Messenger, was particularly influential on the Saddlebred breed). During the nineteenth century, the breed was sometimes known as the Kentucky Saddler, and was occasionally crossed with Morgans, Standardbreds, and Hackneys. The American Saddlebred is known for its five gaits: the walk, the trot, the canter, the slow gait, and the rack. The slow gait and the rack are both four-beat gaits, but they are differentiated by the speed at which the gait is performed, with the slow gait being considerably slower than the more animated rack. Trivia tidbit: Mister Ed, the equine television star of the 1960s, was an American Saddlebred.

Colors: All
Height: 14 to 17hh
Type: Light horse
Gaited: Yes
ALBC Status: None
RBC Status: Endangered
RBST Status: None
Breed Organization Website: www.saddlebred.com or call 859-259-2742

American Standardbred

Breed Description: Developed specifically for harness racing in the mid-1800s, the American Standardbred gets its name from the standard racing speed (a mile in 2 minutes, 30 seconds) that the horses needed to meet in order to be registered. The breed traces directly to two influential foundation sires: the grey Thoroughbred Messenger, and his descendent Rysdick's Hambletonian (also known as Hambletonian 10). It is also significant to note that Rysdick's Hambletonian was a grandson of the Hackney horse Jary's Bellfinder. The combination of the Thoroughbred speed with the driving talent of the Hackney produced the world's top harness racing breed. Even today, the American Standardbred is utilized primarily and extensively for that purpose.

Colors: Typically bay, brown, black, and chestnut. Grey, roan, and tobiano are occasionally seen.
Height: Approximately 15.2hh
Type: Light horse
Gaited: Yes
ALBC Status: None
RBC Status: None
RBST Status: None
Breed Organization Website: www.ustrotting.com

American Warmblood

Breed Description: Ideally, the American Warmblood is similar in type and appearance to the other Warmblood breeds, such as the Swedish, Dutch, or Danish Warmbloods. However, the American Warmblood does not have the lengthy history of which some of the other Warmblood breeds can boast, and therefore the American Warmblood's type is not always as uniform as some of the other breeds. This is compounded by the lack of restriction upon breeds that can crossbreed into the American Warmblood. According to the American Warmblood Society, registration is open to any breed or cross except for purebred Thoroughbred, purebred Arabian, or draft horse breeds. The American Warmblood Society supports the promotion of several disciplines, including dressage, show jumping, eventing, and combined driving.

Colors: All
Height: Approximately 16.1hh, although smaller and larger individuals are often found
Type: Light horse, Warmblood
Gaited: No
ALBC Status: None
RBC Status: None
RBST Status: None
Breed Organization Website: www.americanwarmblood.org or call 501-893-2777

Andalusian

Breed Description: The ancient Andalusian is a Spanish breed, and the Spanish roots are very evident in the Andalusian's muscular build, convex head, and action. The breed is noted for its quiet temperament and gentle nature, as well as its suitability for dressage. The Andalusian has been a major influence in the development of a variety of breeds; some sources say that it has influenced as many as 80 percent of horse breeds worldwide. These include the Norfolk Trotter, the Paso Fino, the Appaloosa, the American Quarter Horse, the Lipizzaner, the Connemara, the Cleveland Bay, and the Welsh Cob.

Colors: Commonly grey, but the International Andalusian and Lusitano Horse Association (IALHA) also recognizes the following colors for purebred Andalusians: bay, black, black bay, brown, chestnut, buckskin, dun, palomino, cremello, perlino, and roan. According to the IALHA, other colors are rare or believed non-existent in the purebred Andalusian.

Height: 15 to 16hh

Type: Light horse

Gaited: No

ALBC Status: None

RBC Status: None

RBST Status: None

Breed Organization Website: www.ialha.org

Anglo-Arabian

Breed Description: In an extremely basic definition, the Anglo-Arabian is an equine with Arabian and Thoroughbred breeding. Occasionally this is achieved by crossing an Arabian with a Thoroughbred, but Anglo-Arabians are also crossed with Anglo-Arabians, or Anglo-Arabians are crossed with Thoroughbreds or even back to purebred Arabians. As long as the resulting foal possesses at least 12.5 percent Arabian blood, the actual cross is of secondary importance. Ideally, the Anglo-Arabian is a quality animal that exhibits the best characteristics of each of the foundation breeds, joining the beauty and elegance of the Arabian with the athleticism and talent of the Thoroughbred. Selective breeding has resulted in exceptional results in the Anglo-Arabians produced in France, and they are particularly well admired for their quality. Lady Wentworth wrote in her 1944 book, *Horses in the Making*, "The best hunter in the world is the half-Thoroughbred half-Arab cross." Anglo-Arab enthusiasts undoubtedly agree with that sentiment!

Colors: Bay, brown, chestnut, or grey
Height: 14.2 to 16.2hh
Type: Light horse
Gaited: No
ALBC Status: None
RBC Status: None
RBST Status: None
Breed Organization Website: www.naaaha.com or you can visit
    www.arabianhorses.org

Appaloosa

Breed Description: Extremely unusual in its color and markings, the Appaloosa has long been a favorite breed in the United States. Distinctive coat patterns are the hallmark of this wonderful stock-type breed. The Appaloosa's name derives from the Palouse River, which is where the Nez Perce Indians initially located these intriguing spotted horses in the 1700s. Of course, the unusual coat patterns immediately attracted a great deal of attention, and the Appaloosa is still making people take notice. The breed's steady temperament and aptitude for Western disciplines have certainly contributed to its popularity, but the fascinating color patterns are undoubtedly a special bonus.

Colors: The Appaloosa Horse Club recognizes the following base colors: bay, brown, black, buckskin, grulla, dun, palomino, cremello, perlino, chestnut, grey, bay roan, blue roan, and red roan. In addition, these base colors may be accompanied by any of seven different coat patterns, including blanket, blanket with spots, roan, roan blanket, roan blanket with spots, and solid.
Other distinctive characteristics: White sclera, striped hooves, mottled skin
Height: 14.2 to 16hh
Type: Light horse
Gaited: No
ALBC Status: None
RBC Status: None
RBST Status: None
Breed Organization Website: www.appaloosa.com

Arabian

Breed Description: Beauty, refinement, strength, stamina, and an unsurpassed history laden with mystic charm—the Arabian is truly a breed without parallel. One must fully utilize the extent of the thesaurus in order to gather enough adjectives to adequately describe the magnificent Arabian. From their delicately dished heads to their gaily-carried tails, the Arabian is truly one of the most beautiful breeds in the world. Although its history is extensive in the Middle East, the Arabian was not largely bred in the United States until the early 1900s. Prior to that time, Arabians in the United States were not common, although George Washington was said to have ridden a half-Arabian, and a son of the Godolphin Arabian was said to have been brought to these shores as early as 1733. The Arabian Horse Registry of America was founded in 1908, and some of the early influential breeders of Arabians in the United States included Homer Davenport, W. K. Kellogg, and Roger Selby. The Arabian has been extremely influential in the development of many other breeds worldwide, from the Haflinger and Thoroughbred to the Percheron and the Welsh Pony. It was also vital to the creation of such breeds as the Morab and the Anglo-Arabian. The Arabian breed is thriving in the United States today, as evidenced by the impressive number of Arabians exhibited at Arabian Horse Association shows each year, where they compete in a wide range of classes. Trivia tidbit: The Arabian's throatlatch is known as the "mitbah."

Colors: Often grey, but also bay, black, and chestnut. "Roan" is also recognized, but this roaning is generally believed to be a result of the sabino or rabicano form of white markings, rather than the traditional roan gene that is seen in American Quarter Horses and Welsh Ponies. The dilute gene does not exist in the Arabian breed, and so there are no purebred palomino, cremello, or buckskin Arabians.

Height: 14.1 to 15.3hh

Type: Light horse

Gaited: No

ALBC Status: None

RBC Status: None

RBST Status: None

Breed Organization Website: www.arabianhorses.org

Belgian

Breed Description: Originating in—you guessed it—Belgium, the Belgian draft horse is known as the *Brabant* in its home country. The breed descended from the Medieval Great Horses, and today the breed is extremely popular for draft hitches. In fact, the Belgian is known as "America's Favorite Draft Horse," according to the Belgian Draft Horse Corporation of America. Their massive size, substance, bones, strength, and muscling make them intimidating and impressive creatures, but their gentle temperaments and innate docility make them a favorite with horse owners. According to Charles S. Plumb's 1919 book, *Judging Farm Animals,* "The Belgian is notable among draft horses for its activity, and the ability to move freely at the trot. Horses of this breed have been raised in Belgium in close touch with the family and so are very docile and easily handled." The Belgian is well known for its signature color: chestnut with a flaxen mane and tail.

Colors: Chestnut (flaxen mane and tail), occasionally other colors
Height: 15.3 to 18hh
Type: Draft horse
Gaited: No
ALBC Status: Recovering
RBC Status: None
RBST Status: None
Breed Organization Website: www.belgiancorp.com

British Riding Pony

Breed Description: *Elegance* is usually the first word that comes to mind when you see a British Riding Pony. These ponies are produced through carefully selected crosses of Thoroughbreds or Arabians with pony breeds. Welsh Ponies are commonly used, as are Connemaras. The ideal result is a talented, beautiful pony with the ability to perform competitively while retaining the inherent beauty for which the native pony breeds are known. The increased size allows for an older child or teenager to continue competing on a larger pony with the same characteristics as their smaller counterparts. Extremely popular in the United Kingdom, the British Riding Pony is gaining in popularity in the United States for use as hunter ponies.

Colors: Usually bay or black; occasionally chestnut, grey, or other solid colors
Height: 12.2 to 14.2hh
Type: Pony
Gaited: No
ALBC Status: None
RBC Status: None
RBST Status: None
Breed Organization Website: www.nationalponysociety.org.uk

Buckskin

Breed Description: Long recognized as hardy, durable, and tough, the Buckskin is favored as a working horse. While it is not technically a breed, the International Buckskin Horse Association (IBHA) was established in 1971 in order to register buckskins and duns of many breeds. The breed's striking color (a cream-colored body accompanied by a jet-black mane, tail, and points) has made it a perennial favorite with horse enthusiasts. Although a wide variety

of breeds are recognized by the IBHA, the majority of Buckskin registrations are for stock-type horses, possibly stemming from the fact that stock-type horses are extremely popular in any case, but also possibly from the fact that the buckskin coloring appears more often in the stock-type breeds than it does in the traditionally English breeds. One important aspect to understand is the genetic difference between the colors buckskin and dun. The simple action of a dilute (cream) gene changes a horse that would have ordinarily been bay and lightens the color to buckskin. Dun is an entirely different color gene and is commonly seen in American Quarter Horses. Dun is not controlled by the dilute (cream) gene, and there is actually an entire range of dun-related colors and shades. A very basic rule of thumb (and not always 100 percent accurate) is that duns have a dorsal stripe and leg barring, while buckskins do not. While there are obviously exceptions to both cases, this statement is generally true. Trivia tidbit: Dale Evans' horse, Buttermilk, was a Buckskin.

Colors: Buckskin
Height: An average of 15hh
Type: Light horse
Gaited: Occasionally
ALBC Status: None
RBC Status: None
RBST Status: None
Breed Organization Website: www.ibha.net

Canadian

Breed Description: Developed in the late seventeenth century from French foundation stock, the Canadian breed has faced extinction on more than one occasion despite the fact that the breed once existed in large numbers. Great numbers of Canadian horses were exported to New England during the 1800s, and these horses served as the foundation animals for several breeds including the Morgan, the Saddlebred, and the Standardbred. The Societe des Eleveurs de Chevaux Canadiens (The Canadian Horse Breeders Association) was established in 1895 and today has a membership of more than 1,100. The Canadian is noted for its longevity, as well as its resemblance to the Morgan horse. In fact, it has been suggested that Justin Morgan, the foundation sire of the Morgan breed, was actually a Canadian horse. The breed has often been referred to as the "Little Iron Horse."

Colors: Mainly bay and black, but also chestnut and brown
Height: 14 to 16hh
Type: Light horse
Gaited: No
ALBC Status: Threatened
RBC Status: At risk
RBST Status: None
Breed Organization Website: www.lechevalcanadien.ca

Caspian

Breed Description: While many breeds lay claim to the title of oldest breed, the Caspian breed is generally accepted as being one of the very oldest, if not the oldest breed in existence today. Despite this long and far-reaching pedigree, the Caspian was nearly extinct in the 1960s, and it was only through the efforts of a devoted Caspian enthusiast, Louise Firouiz, that the breed has been preserved and perpetuated. Many believe that the Caspian is the ancestral breed of the Arabian, and there are several characteristic similarities between the two breeds, despite the disparity in their sizes. Genetic testing has shown a close connection between the Caspian and the Arabian breeds. Although they are still extremely rare in the United States, one of the top sires in the Hunter Pony Breeding division of the United States Equestrian Federation is a Caspian.

Colors: Often bay, grey, and chestnut; sometimes black and dun. White markings are typically minimal.
Height: Average of 11.2hh, ranging from 10 to 12hh
Type: Pony-sized, but is actually a small light horse breed
Gaited: No
ALBC Status: Critical
RBC Status: None
RBST Status: None
Breed Organization Website: www.caspian.org

Chincoteague

Breed Description: *Misty of Chincoteague*, the classic *Newbery* Honor book by Marguerite Henry, has brought immeasurable recognition to the Chincoteague breed since its publication in 1947. The Chincoteague and Assateague islands in the eastern United States are the home of the Chincoteague pony breed. Legend has it that the ancestors of today's wild ponies were shipwrecked on the islands centuries ago, but this legend has never been proven as fact. Today the ponies of Chincoteague vary widely in type and appearance due to the indiscriminant breeding and the wide range of breeds that have been infused into the island ponies over time. These breeds have ranged from the Thoroughbred to the Arabian to the Mustang, as well as other pony breeds. Arabian breeding was infused on more than one occasion in hopes of improving the overall quality of

the island ponies. Each year, the Chincoteague Island Fire Department hosts the Pony Penning, during which the ponies are brought from the island to the mainland and some are sold at auction. While they are known as ponies due to their diminutive size, the Chincoteague's leg-to-body ratio is technically more horse-like in appearance than pony-like.

Colors: All, including pinto patterns
Height: Their heights range from 12 to 14.2hh, with many averaging about 13.2hh
Type: Pony
Gaited: No
ALBC Status: None
RBC Status: None
RBST Status: None
Breed Organization Website: www.pony-chincoteague.com

Cleveland Bay

Breed Description: An ancient breed with a history dating back to the mid-1700s, the Cleveland Bay was originally known as the Chapman. The name was later changed to the Cleveland Bay in reference to the area of England from which the breed originated. While the Cleveland Bay is sometimes referred to as Britain's oldest equine breed, it has become very rare in recent years. Because the Cleveland Bay's prime use was as a coach or carriage horse, the advent of the automobile served to diminish the importance of the breed, and its population has decreased significantly. The breed has long been noted for its bay coloring without white markings, but the registration restrictions on white markings have been relaxed in recent years in order to help preserve this critically endangered breed. The Cleveland Bay is the only horse or pony breed that is currently listed as critical on the lists of all three of the breed conservation organizations.

Colors: Bay, without white markings
Height: 16 to 16.2hh
Type: Light horse
Gaited: No
ALBC Status: Critical
RBC Status: Critical
RBST Status: Critical
Breed Organization Website: www.clevelandbay.org

Clydesdale

Breed Description: The Clydesdale is a beautiful draft breed that originally hailed from the area of Clyde in Scotland, from which its name derives. The breed was developed during the 1700s, and the foundation breeds included the Belgian and the Shire, along with native draft horses. Despite its immense size and substance, the Clydesdale is noted for its attractive head, accentuated by a broad forehead. The breed is also noted for its impressive white markings: wide blazes and high stockings. The Clydesdale is a popular choice with draft driving hitches in the United States, with the most well known undoubtedly being the Budweiser Clydesdales. The Budweiser Clydesdales have truly been beneficial in raising interest and enthusiasm for the draft breeds in general and the Clydesdale breed in particular.

Colors: Bay, brown, or black; rarely chestnut. Extensive white markings on the face and legs are encouraged.
Height: The ideal height is approximately 17hh, but they range from 16.2 to 18hh
Type: Draft horse
Gaited: No
ALBC Status: Watch
RBC Status: At risk
RBST Status: Vulnerable
Breed Organization Website: www.clydesusa.com

Colonial Spanish Horse (Mustang)

Breed Description: The Colonial Spanish Horse is not considered a single specific breed, but instead is a name that refers to several directly linked breeds that share the common ancestry of the original Spanish horses that were brought to North America by the explorers of the 1500s. These breeds include the Spanish Barb, the Spanish Mustang, the Kiger Mustang, the Pryor Mountain Mustang, the Marsh Tacky, the Abaco Barb, the Florida Cracker Horse, and the Choctaw, among others. The Colonial Spanish Horse strains are listed as critical on the American Livestock Breeds Conservancy watch list, and the various strains exhibit many similar physical characteristics, not the least of which is their small size (usually less than 15hh). They are notably good-tempered equines with a fascinating heritage.

Colors: Virtually any color (chestnut, black, bay, dun, buckskin, grullo, red dun, cream, and palomino), as well as grey and roan. Pinto and Appaloosa markings are also common.
Height: 13.2 to 15hh
Type: Light horse
Gaited: Occasionally
ALBC Status: Critical
RBC Status: None
RBST Status: None
Breed Organization Website: www.horseoftheamericas.com

Colorado Ranger

Breed Description: At first glance, the Colorado Ranger bears a striking resemblance to the Appaloosa breed, but in actuality, they are recognized as two separate breeds. The Colorado Ranger was developed in Colorado, heavily based on Arabian and Barb breeding in the early days. In the early 1900s, a man named Mike Ruby began working with some unusually marked Colorado horses, and it was his foundation stallions (Max #2 and Patches #1) that truly established the breed. Even today, the requirement for registration in the Colorado Ranger Horse Association (CRHA) is that the horse must trace back to either Max or Patches. Mike Ruby kept meticulous records, and this history has proved important, particularly considering the fact that he founded the CRHA in the 1930s. It is estimated that one in eight registered Appaloosas would qualify for registration with the CRHA, and it is estimated that 90 percent of Colorado Rangers are double registered with the Appaloosa Horse Club.

Colors: All colors are eligible for registration, regardless of the presence of Appaloosa markings

Height: 14.2 to 16hh

Type: Light horse

Gaited: No

ALBC Status: None

RBC Status: None

RBST Status: None

Breed Organization Website: www.coloradoranger.com

Connemara

Breed Description: The Irish example of a British native pony, the Connemara shares many of the characteristics of the other native pony breeds, including an attractive head, although the Connemara's head is not usually as refined as that of the Welsh Pony. Infusions of Arabian, Thoroughbred, and Welsh Cob breeding have improved the overall quality of the original native Connemaras, and today's ponies are very attractive. The Connemara is noted for its talent, athleticism, and hardiness, and its larger size (medium to large pony size) is an important attribute for many families, particularly if their child has outgrown their smaller Welsh or Shetland pony. There is lesser emphasis on knee action in the Connemara's movement than there is in the Welsh breeds, but the Connemara's depth and substance are very important.

Colors: Grey, bay, brown, dun, chestnut, roan, black, and palomino
Height: 13 to 15hh
Type: Pony
Gaited: No
ALBC Status: None
RBC Status: None
RBST Status: None
Breed Organization Website: www.acps.org

Dales

Breed Description: The Dales pony is another of the British native pony breeds and is extremely rare in the United States. Unlike the flatter movement of the Connemara pony, the Dales pony is expected to make good use of its knees and hocks with good action at the trot. Dales ponies are known for their quality legs and feet, and it has been said that they have the best legs of any of the British native pony breeds. Their heads lack the concave, dished profile that the Welsh are noted for, but the Dales ponies are known for their "tippy" ears with curved tips. As is the case with many of the British native pony breeds, the Dales pony is shown naturally without the trimming of hair. The Dales pony stands a substantial 13.2 to 14.2hh, which makes it an excellent size for a family pony.

Colors: Mainly black, brown, grey, or bay. Rarely roan, never pinto or chestnut. Minimal white markings are allowed.
Height: 13.2 to 14.2hh
Type: Pony
Gaited: No
ALBC Status: Study
RBC Status: Critical
RBST Status: Endangered
Breed Organization Website: www.dalesponies.com

Danish Warmblood

Breed Description: The Danish Warmblood is one of the newer Warmblood varieties, having been developed in Denmark during the 1960s. The Danish Warmblood registry was established in 1962, and today the Danish Warmblood is noted for its superior conformation. The foundation breeds of the Danish Warmblood included the Frederiksborg and Holsteiner, along with the Thoroughbred and the Trakehner. The Hanoverian is absent from the list of foundation breeds, and the lack of Hanoverian blood has lent a unique look to the Danish Warmbloods, as opposed to the majority of the other Warmblood breeds that do exhibit Hanoverian influence. The North American Danish Warmblood Association was founded in 2001 in order to promote the *Dansk Varmblod* in North America.

Colors: Commonly bay, but any solid color is acceptable
Height: Approximately 16.2hh
Type: Light horse, Warmblood
Gaited: No
ALBC Status: None
RBC Status: None
RBST Status: None
Breed Organization Website: www.danishwarmblood.org

Dartmoor

Breed Description: The Dartmoor is among the smallest of the British native pony breeds. Typically found in very dark colors, such as bay or black, the lighter-colored Dartmoors (chestnut and grey) are considerably less popular and were even frowned upon in the show ring years ago. According to Alex Fell's book, *The Complete Book of In-Hand Showing,* "There was agreement over the features of the breed, though, like the importance of small ears. They didn't like chestnuts . . . or a pony with a white foot." The Dartmoor exhibits many of the characteristics for which the British native ponies are known, including hardiness and a small head. As the Dartmoor breed standard states, "The Dartmoor is a very good-looking riding pony, sturdily built yet with quality."

Colors: Bay, black, brown, chestnut, and roan. Occasionally grey. Pinto markings are not accepted, and white markings in general are not encouraged.
Height: From 11 to 12.2hh
Type: Pony
Gaited: No
ALBC Status: Threatened
RBC Status: Critical
RBST Status: Vulnerable
Breed Organization Website: www.dartmoorpony.com

Donkey

Breed Description: Donkeys are a truly unique member of the equine family. Donkeys are notably different from their horse cousins in a number of ways, not the least of which is the generous length of the Donkey's ears. In addition to this, the Donkey also exhibits primitive markings, which include a shoulder cross, dorsal stripes, and leg barring. Donkeys are also known for their sparseness of mane and tail hair. Donkeys can be found in a rainbow of colors and a wide range of sizes, from the petite (and adorable) Miniature Mediterranean Donkey to the impressive and substantial Mammoth Stock Donkey. It is important to remember that Donkeys are not referred to as stallions and mares; rather, the male Donkey is known as a Jack, and the female Donkey is called a Jenny or Jennet. Trivia tidbit: There is a legend behind the Donkey's shoulder cross marking, and it is said that because it was a Donkey that Mary rode into Bethlehem on the night Jesus Christ was born, Donkeys are forever marked with the special shoulder cross.

Colors: Grey, roan, cream, black
Heights: Miniature Mediterranean Donkey: Less than 36 inches
Standard Donkey: 36 to 48 inches
Large Standard Donkey: 48 to 56 inches (48 to 54 inches for females)
Mammoth Stock Donkeys: Over 54 inches (females); over 56 inches (males)
Type: Donkey
Gaited: No
ALBC Status: The Miniature Donkey and the American Mammoth Jack are both listed as critical
RBC Status: None
RBST Status: None
Breed Organization Website: www.lovelongears.com

Dutch Warmblood

Breed Description: The athletic and talented Dutch Warmblood is well known for its success in dressage and its excellent ability over fences. This success may be due in part to the fact that the Dutch Warmblood was developed with an emphasis on enhancing performance ability rather than focusing on beauty. The Dutch Warmblood descends from two native Dutch breeds: the lighter Gelderlander and the heavier Gronigen. These native breeds were crossed with Thoroughbreds, and the resulting Dutch Warmblood is often a lighter type of Warmblood than some of the other breeds. For many years, Dutch Warmbloods have been produced with three different goals in mind, and as such, three different types have evolved; these include the Riding type, the Harness type, and the Gelders type (an all-around horse). Subsequently, three different versions of the Riding type are recognized in North America—dressage, jumping, and hunter.

Colors: Commonly bay, but also grey, black, brown, and chestnut. Roan and tobiano are also occasionally seen.
Height: 15.2 to 17hh, an average of 16.2hh
Type: Light horse, Warmblood
Gaited: No
ALBC Status: None
RBC Status: None
RBST Status: None
Breed Organization Website: www.kwpn.nl

Eriskay Pony

Breed Description: Nearly everyone is familiar with the Shetland pony breed, and many people are familiar with the Highland pony breed, but very few people are aware of the existence of a third Scottish pony breed, the Eriskay. "The rarest of the rare" is probably an apt description of the highly endangered Eriskay pony. In fact, it is one of only three breeds listed as critical on the Rare Breeds Survival Trust (UK) conservation list. The breed was nearing extinction (with less than two dozen Eriskay ponies in existence) in the 1970s before measures were taken to protect and preserve this native breed. The Eriskay Pony Society was established in 1995 to promote "these rare ponies of remarkable temperament and hardiness [that] are carving a special niche for themselves in today's competitive world," according to the society's statement.

Colors: Commonly grey, occasionally bay or black. According to the Eriskay Pony Society, these are the only colors that occur in the Eriskay breed.
Height: 12 to 13.2hh
Type: Pony
Gaited: No
ALBC Status: None
RBC Status: None
RBST Status: Critical
Breed Organization Website: www.eriskaypony.com

Exmoor

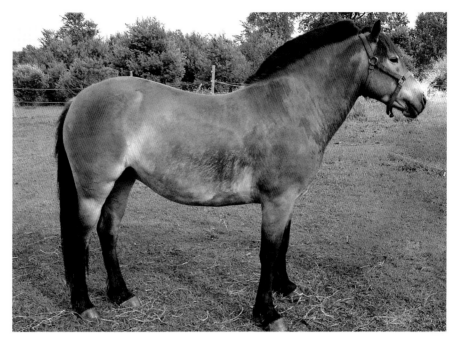

Breed Description: Extremely rare in the United States, the ancient Exmoor breed has quietly existed on the moors of England for time immemorial. Slightly larger than the Dartmoor, the Exmoor is known for being a bit plainer through the head than the Dartmoor. Unique to the Exmoor is its unusual two-layered coat that is especially suited for repelling water and protecting the pony throughout harsh winter conditions. The Exmoor is very distinctive and easily recognizable due to its consistently dark coloring, brown or bay, with a pale muzzle. Resilient and enduring, the Exmoor pony is one of Britain's equine treasures.

Colors: Brown and bay, with lighter colored muzzle ("mealy muzzle," said to be the color of oatmeal). Also occasionally a dark dun. White markings are prohibited in any amount.
Height: Up to 12.3hh (stallions and geldings), up to 12.2hh (mares)
Type: Pony
Gaited: No
ALBC Status: Critical
RBC Status: Critical
RBST Status: Endangered
Breed Organization Website: www.exmoorponysociety.org.uk

Fell

Breed Description: The Fell pony is a British native pony breed that hails from England, as does its close cousin, the Dales pony. Both breeds are extremely rare in the United States, but the Fell pony is easily recognizable due to its distinctive black coloring, as well as for its masses of feathering. The foundation breeds of the Fell pony include the Galloway and the Friesian. The connection to the Friesian is particularly notable, as the Fell pony distinctly resembles a miniature Friesian. Like many of the native pony breeds, the Fell is characterized by its impressive trotting ability, substance and bone, and presence. Dedicated enthusiasts are working to increase awareness in America about this fascinating native pony breed.

Colors: Black; occasionally chestnut; rarely bay, brown, black, or grey
Height: 13 to 14hh
Type: Pony
Gaited: No
ALBC Status: Threatened
RBC Status: Critical
RBST Status: At Risk
Breed Organization Website: www.fellpony.org

Foundation Quarter Horse

Breed Description: Because of the overwhelming popularity of the American Quarter Horse, the popularity of the Foundation Quarter Horse is not surprising. The National Foundation Quarter Horse Association, the Foundation Quarter Horse Association, the Foundation Horse Registry, the Foundation Quarter Horse Registry, and the Foundation Quarter Horse Breeders Association are organizations that have been founded with the express purpose of emphasizing the original visions of early American Quarter Horse breeder Robert Denhardt. The Foundation Horse Registry states that its purpose is to "preserve and protect the original Quarter Horse, and immortalize for generations to come, the legendary bloodlines." In essence, these organizations seek to avoid Quarter Horses with high percentages of Thoroughbred blood, in hopes of preserving the original Quarter Horse type, including a deep body, heavy muscling, and the breed's inherent cow sense. The ideal Foundation Quarter Horse contains less than 20 percent Thoroughbred blood, and less than 10 percent is considered even better.

Colors: sixteen colors are recognized: brown, black, grey, sorrel, bay, chestnut, dun, red dun, grullo, buckskin, perlino, cremello, bay roan, blue roan, red roan, and palomino. The blue roan and grullo are particularly popular with Foundation Quarter Horse enthusiasts.
Height: 14.3 to 16hh
Type: Light horse
Gaited: No
ALBC Status: None
RBC Status: None
RBST Status: None
Breed Organization Website: www.fqha.com, www.foundationhorses.com, www.fqhr.net, www.bhfqh.com, www.nfqha.com

French Saddle Pony

Breed Description: The *Poney Francais de Selle* (which translates to "the French Saddle Pony") is a talented breed with the athletic ability to compete in a variety of disciplines from dressage to driving. The French Saddle Pony was developed in France and is comparable in type and quality to the German Riding Ponies and British Riding Ponies. The French Saddle Pony is produced through crosses similar to its German and British counterparts—native pony breeds such as Welsh and New Forest crossed with horse breeds such as the Selle Francais or the Arabian. While still relatively rare in the United States, the high interest in the German and British Riding ponies makes it very likely that the French Saddle Pony will increase in popularity in years to come.

Colors: All
Height: 12.1 to 14.2hh
Type: Pony
Gaited: No
ALBC Status: None
RBC Status: None
RBST Status: None
Breed Organization Website: None

Friesian

Breed Description: Quickly gaining in popularity in the United States, the Friesian is a regal, noble breed that is easily recognizable due to its black coloring and its abundance of feathering on the fetlocks. The Forest Horse of the Netherlands influenced the breed early on, and the result was a massive, baroque-type horse. The Friesian went through a period where a lighter, more refined version was popular, but today's substantial Friesians are more consistent with the original breed type. It has been said that all Friesians trace back to a single foundation stallion, Nemo 51. The registry opened in 1879 in Holland, and the Friesian Horse Association of North America was founded in the mid-1980s. An inspection process helps to ensure the quality and type of the Friesian breed, with 60 percent of the inspection score being weighted on the Friesian's movement (the remaining 40 percent is based on the horse's conformation). The Friesian's awe-inspiring brilliance, coupled with its impressive movement and dramatic appearance in harness or under saddle, makes it no surprise that the breed is enjoying a surge of popularity in the United States.

Colors: Black is the only recognized color, although chestnuts do occasionally occur. White markings are prohibited, with the exception of a small star.

Height: 14.3 to 16hh (15.3 is considered ideal by the Friesian Horse Association of North America). For registration, stallions must stand at least 15.3hh by four years of age; mares and geldings must stand at least 14.3hh.

Type: Light horse

Gaited: No

ALBC Status: Recovering

RBC Status: None

RBST Status: None

Breed Organization Website: www.fhana.com or www.friesianhorsesociety.com

German Riding Pony

Breed Description: Ideally more like a miniature Warmblood than a pony, the German Riding Pony (or *Deutsche Reitpony*, as it is known in Germany) is an attractive breed that was developed in the 1960s. As is the case with the British Riding Pony, the German Riding Pony was founded by crossing native pony breeds (Welsh, Connemara, New Forest, etc.) with horse breeds such as the Arabian, Anglo-Arabian, and Thoroughbred. In addition, the development of the German Riding Pony also included infusions of Hanoverian, Holsteiner, Norwegian Fjord, and Haflinger, but these breeds were not considered to produce satisfactory results. Breeders of German Riding Ponies have struggled with a lack of consistency in breed type, and outcrosses to Welsh and Arabian are still occasionally

used. In addition, due to the rarity of the German Riding Pony in the United States, the registry has been expanded to include a variety of bloodlines and breeds. For instance, purebred Welsh Ponies have been inspected and approved by the Weser-Ems registry (the United States registry for German Riding Ponies) in an attempt to increase and broaden the German Riding Pony gene pool in the United States.

Colors: Commonly bay and black
Height: 13.2 to 14.2hh
Type: Pony
Gaited: No
ALBC Status: None
RBC Status: None
RBST Status: None
Breed Organization Website: www.oldenburghorse.com/Oldbrg-Ponies.htm

Gotland

Breed Description: The Swedish contribution to pony breeding, the Gotland pony shares many of the characteristics that are prized in the British native pony breeds. Hardy, resilient, enduring, and industrious, the breed is known for its calm demeanor and willingness to work. The Gotland pony is extremely rare in the United States, and the population has only recently begun to increase. There are a handful of dedicated Gotland breeders who are striving to preserve and promote the Gotland pony in America, and they are utilizing their ponies in harness and under saddle, demonstrating the excellent characteristics of this very rare breed.

Colors: Commonly bay or black, although the Gotland can be any color except for dun, grey, or pinto
Height: Up to 13hh
Type: Pony
Gaited: No
ALBC Status: Watch
RBC Status: None
RBST Status: None
Breed Organization Website: www.gotlandponies.org

Gypsy Vanner

Breed Description: For decades, these small Irish draft horses have existed as caravan horses for the Gypsies in the United Kingdom. Their background includes Clydesdale, Friesian, and Shire, as well as native pony breeding. The first Gypsy horses were brought to the United States in the 1990s, and they have exploded in popularity in the United States since that time. Their docile temperaments have contributed to their popularity, but the intriguing charm of their history as Gypsy caravan horses is also part of what makes them desirable to so many people. The Gypsy Vanner is noted for its heavy bones and abundant feathering, as well as its strength. There are several American registries for Gypsy horses, and the breed is often referred to by a variety of names, including the Gypsy Vanner, the Gypsy Cob, the Irish Tinker, the Tinker, and the Vanner.

Colors: All, but they are particularly noted for their pinto patterns
Height: Any height is permissible, but 14.2 to 16hh is most common
Type: Draft
Gaited: No
ALBC Status: None
RBC Status: None
RBST Status: None
Breed Organization Website: www.gypsyvannerhorse.org, www.aghba.org, or www.gypsycobsociety.com

Hackney Horse

Breed Description: As is the case with the related Hackney Pony, the Hackney Horse is an English breed that is well known as a driving animal. Its extravagant action, characterized by impressive knee and hock action, makes it a splendid choice for use in harness. The Hackney Horse was developed during the late eighteenth and nineteenth centuries, bred from Norfolk Trotters and Thoroughbreds, and the name is derived from the French word *haquenee* which means "nag" or "gelding." The Hackney stud book was formed in 1883, but the advent of automobiles significantly reduced the need for horses in harness, and thus the Hackney population has seriously decreased.

Colors: Black, bay, brown, and chestnut. White markings are permissible.
Height: 14.2 to 16.2hh
Type: Light horse
Gaited: No
ALBC Status: Critical
RBC Status: Endangered
RBST Status: Endangered
Breed Organization Website: www.hackneysociety.com

Hackney Pony

Breed Description: Harness pony extraordinaire! Its high tail carriage accentuates the Hackney Pony's highly animated trot. The breed possesses definite pony character and is not simply a Hackney Horse in miniature. The breed was developed in England during the late nineteenth century by crossing native pony breeds such as Fell or Welsh with the Hackney Horse. As you may recall, the Hackney Horse was descended from the Norfolk Trotter, an ancient breed that was of Andalusian, Spanish, and Arabian breeding. The toes of the Hackney Pony are typically allowed to grow long in an attempt to accentuate the action. One of the best descriptions of Hackney movement appears in Herbert H. Reese's 1958 book, *The Kellogg Arabians*: " 'High' action for an Arab brings the knee as high as the elbow. Hackney—and show ponies—bring the *hoof* as high as the elbow, and some Saddlebreds go almost as high."

Colors: Bay, black, brown, and chestnut, occasionally grey or roan
Height: 12 to 14hh
Type: Pony
Gaited: No
ALBC Status: Although the Hackney Horse is listed as critical on the ALBC list, the Hackney Pony is not considered to be threatened and therefore is not listed
RBC Status: Endangered
RBST Status: Endangered
Breed Organization Website: www.hackneysociety.com

Haflinger

Breed Description: A native of Austria, the Haflinger is an attractive breed that is highly recognizable due to its striking chestnut coloring. The breed's characteristic flaxen mane and tail accentuate the chestnut color. Haflingers are well known for their gentle natures, their longevity, and their natural hardiness. Bred from Arabian foundation stock that was crossed with native pony breeds, the Haflinger has developed into a substantial, muscular breed with ample bone and strength. The Arabian and native pony influence has also contributed to the Haflinger's attractive face. Haflingers are popular driving ponies in the United States and are also excellent family ponies.

Colors: Rich chestnut with flaxen mane and tail. White face markings are common and encouraged, although white leg markings are discouraged.
Height: 13.2 to 15hh. Smaller individuals are discouraged from use as breeding animals, but larger individuals are allowed if the horse is a particularly quality example.
Type: Pony
Gaited: No
ALBC Status: None
RBC Status: None
RBST Status: None
Breed Organization Website: www.haflingerhorse.com

Hanoverian

Breed Description: The Hanoverian is a Warmblood breed that was developed in northern Germany during the eighteenth century. The breed was named for the kingdom of Hannover, where it originated, and was established from a foundation of several breeds, including the Holsteiner, the Thoroughbred, and the Trakehner. The State Stud in Celle, Germany, was established in 1735, and centuries of selective breeding have resulted in the Hanoverian's superior excellence. Today's Hanoverian is well admired for its talent as a show jumper and its aptitude for dressage. The Hanoverian has also proven to exhibit the athleticism and talent necessary to succeed at the Olympic level.

Colors: Chestnut, bay, brown, black, and grey. Dilutes, such as the palomino, buckskin, and cremello are excluded, as are horses with excessive white markings.
Height: 15.3 to 17hh
Type: Light horse, Warmblood
Gaited: No
ALBC Status: None
RBC Status: None
RBST Status: None
Breed Organization Website: www.hanoverian.org

Highland Pony

Breed Description: The Highland pony, which is named for its home on the Scottish Highlands, was historically utilized by Scottish crofters and farmers. Today its more common use is as a family pony. Some sources suggest a historic connection between the Highland ponies and the Norwegian Fjord horse, owing to the fact that Highland ponies often exhibit dun coloring and primitive markings (including a dorsal stripe and zebra markings). The Highland pony breed profile lists ten different colors in which the Highland pony is recognized. Five of these colors are shades of dun. White markings are heavily discouraged, and consequently, today's Highland ponies typically exhibit only dark hooves. The Highland Pony Society was formed in 1923.

> **Colors:** Commonly a shade of dun, including mouse, yellow, golden/grey, cream, or fox. Also commonly grey. Occasionally black, brown, bay, or liver chestnut. White markings are discouraged.
> **Height:** 13 to 14.2hh
> **Type:** Pony
> **Gaited:** No
> **ALBC Status:** None
> **RBC Status:** None
> **RBST Status:** At Risk
> **Breed Organization Website:** www.highlandponysociety.com

Holsteiner

Breed Description: Like the Hanoverian, the Holsteiner is a Warmblood of German origin; however, the development of the Holsteiner predates that of the Hanoverian by several centuries. The American Holsteiner Association states that the Holsteiner's history extends back 700 years, whereas the development of the Hanoverian was not undertaken until 1735. The Holsteiner's name derives from the Schleswig-Holstein region of Germany, and the breed was influenced over time by other breeds, such as the Cleveland Bay, the Thoroughbred, the Anglo-Arabian, and the Selle Francais. As is the case with the Hanoverian, the Holsteiner's talent and athleticism have made it a perennial choice for equestrian sports at the highest levels.

Colors: Brown, black, grey, bay, and chestnut. Pinto and Appaloosa patterns are not acceptable, and dilute colors, such as buckskin or palomino, are also discouraged.
Height: 16 to 17hh
Type: Light horse, Warmblood
Gaited: No
ALBC Status: None
RBC Status: None
RBST Status: None
Breed Organization Website: www.holsteiner.com

Hunter Pony

Breed Description: As is the case with the sport pony, the hunter pony is not a breed per se, and there is no registry or stud book specifically maintained for hunter ponies. However, the United States Equestrian Federation (USEF) does maintain records for hunter ponies and calculates points for the leading sires of Hunter Ponies in both the Hunter Pony breeding (in-hand) division, as well as the Hunter Pony performance divisions. While a hunter pony can be of any breed as long as it is under 14.2hh, there is a preponderance of Welsh Pony blood in the majority of hunter ponies in the United States today. In fact, eleven of the top fifteen current hunter pony breeding sires are Welsh; two are crossbred Welsh; one is a Hanoverian, and one is a Caspian. In the Hunter Pony performance category, fourteen of the fifteen leading sires are Welsh or Welsh-cross, and one is a Connemara. In order to be competitive in the Hunter Pony divisions, the ponies must be excellent jumpers with good form over fences, as well as fluid movers with long strides.

Colors: All
Height: Up to 12.2hh (small division), up to 13.2hh (medium division), up to 14.2hh (large division)
Type: Pony
Gaited: No
ALBC Status: None
RBC Status: None
RBST Status: None
Breed Organization Website: www.usef.org offers excellent information about the Hunter Pony divisions

Icelandic

Breed Description: Rugged and tough, the Icelandic horse was bred in Iceland for nearly 1,000 years without outside influence. While Icelandics have been exported to many other countries, no horses are allowed to be imported into Iceland. Icelandic breeders have also maintained an extensive inspection process for their breeding stock. The inspections take place when the horse is five years old, and the inspections have a 40 percent emphasis on conformation and a 60 percent emphasis on riding quality. The Icelandic is a gaited breed with five gaits: the walk, the trot, the canter, the tölt (a type of running walk), and the flying pace (also called the flugskeið). Icelandics are notably long-lived, often reaching their 30s, and occasionally living to over 40 years old.

Colors: All, including pinto patterns
Height: 12 to 14hh
Type: Light horse
Gaited: Yes
ALBC Status: None
RBC Status: None
RBST Status: None
Breed Organization Website: www.ihsgb.co.uk

Irish Draught

Breed Description: Although the name indicates otherwise, the Irish Draught is not a draft horse in the traditional sense. It is an ancient Irish breed that has an extensive history as a workhorse, but is not of draft horse breeding. Instead, it was founded on such breeds as the Flanders and the Spanish breeds. The official breed profile for the Irish Draught calls for "an active short-shinned powerful horse with substance and quality . . . exceptionally strong and sound constitution." The breed is very athletic and talented with excellent jumping ability. They are noted for being good-natured family horses and are the foundation breed for the Irish Hunter, which is an Irish Draught/Thoroughbred cross. The Irish Hunters are very popular as sport horses.

Colors: Commonly black and grey, but other solid colors are permitted. Extensive white markings are discouraged.
Height: 15.1 to 16.3hh
Type: Light horse
Gaited: No
ALBC Status: Study
RBC Status: Critical
RBST Status: None
Breed Organization Website: www.irishdraught.com

Lac La Croix Indian Pony

Breed Description: A relative newcomer in terms of horse breeds, the Lac La Croix Indian Pony (LLCIP) is actually a very old breed that has only recently received recognition in its own right. The breed hails from the northern region of Minnesota and northern Ontario, where these small Indian Ponies have existed for centuries. It has only been in recent years that the breed has been officially named the Lac La Croix Indian Pony and preservation has been undertaken by the Rare Breeds Canada organization. The breed is still extremely rare, but is thankfully past any danger of extinction. Genetic testing has revealed a background of Iberian (Spanish horse) ancestry, along with native pony breeding. Some sources believe that the Canadian horse is also an ancestor of the LLCIP.

Colors: Commonly found in dark colors such as bay or black, as well as dun or grullo. The cream gene (which is the mechanism behind palomino, buckskin, and smoky black) is absent from the Lac La Croix Indian Pony breed. Face and leg markings are acceptable except for above the knee, and pinto or Appaloosa patterns are not permitted.

Height: 12 to 13.2hh

Type: Pony

Gaited: No

ALBC Status: None

RBC Status: Critical

RBST Status: None

Breed Organization Website: www.rarebreedscanada.ca or www.equilore.ca

Latvian

Breed Description: Named for its country of origin, this Warmblood type of light horse was developed during the first half of the twentieth century. The breed was greatly influenced by Oldenburg and Hanoverian bloodlines and is now considered to have evolved into three types: the Latvian Draft, the Latvian Harness horse, and the Lightweight Latvian (more influenced by Thoroughbred and Hanoverian bloodlines). Latvians are known for their pleasant demeanor and are inspected and registered in the United States by the Rheinland Pfalz-Saar organization.

Colors: Bay, brown, black, grey, and chestnut
Height: 15.1 to 16hh
Type: Light horse, Warmblood
Gaited: No
ALBC Status: None
RBC Status: None
RBST Status: None
Breed Organization Website: None

Lipizzan (Lipizzaner)

Breed Description: If you think of the Lipizzan, it is quite likely that you automatically think of their wonderful dressage talent and the skill of the Lipizzan stallions that perform at the World Famous Lipizzaner Stallions tour. It is estimated that more than 23 million people have had the opportunity to observe the marvelous Lipizzan stallions. The breed owes much to its six foundation stallions from the 1700s and 1800s. They included Kladrub, Neapolitan, and Arabian stallions that were crossed with Spanish horses (often Andalusian). The Lipizzan (known as the Lipizzaner in Europe) is the national symbol of Slovenia, as it is the only breed of horse to have been developed in that country.

Colors: Nearly always grey, very rarely bay or black
Height: 15 to 16.1hh
Type: Light horse
Gaited: No
ALBC Status: Threatened
RBC Status: None
RBST Status: None
Breed Organization Website: www.lipizzan.org

Miniature Horse

Breed Description: Delightful and diminutive, the Miniature Horse is ideally more horse-like than pony-like, despite their size. Their leg-to-body ratio should be closer to that of a horse than a pony. The aim of Miniature Horse breeders is to strive for a Mini that perfectly resembles a full-size horse in all respects except for size. The foundation breeds of the Miniature Horse included the Falabella, the Shetland, the Hackney, the Pony of the Americas, and the Welsh Pony. The first registry for Miniature Horses was established in 1972 as a division of the Shetland Pony Club. This registry (the American Miniature Horse Registry) recognizes two divisions of Minis: Division A, which registers Minis that are 34 inches and under; and Division B, which registers Minis that are between 34 and 38 inches. A second registry, the American Miniature Horse Association, was founded in 1978 and registers Miniature Horses that are 34 inches and under when measured at the last hair of their mane. Miniature Horses are popular for driving and also are popular with equine enthusiasts who seek the joy of owning a horse in a tiny package.

Colors: All colors or patterns, including pinto patterns. Any eye color is also acceptable.
Height: Under 38 inches
Type: Pony
Gaited: No
ALBC Status: None
RBC Status: None
RBST Status: None
Breed Organization Website: www.shetlandminiature.com or call 309-263-4044

Missouri Fox Trotting Horse

Breed Description: Recognized by some as the ultimate trail and pleasure riding breed, the Missouri Fox Trotting Horse is a gaited breed that was (not surprisingly) developed in Missouri. This sure-footed breed is well suited to working on rough and uneven terrain. The Missouri Fox Trotting Horse Breeders Association estimates that 90 percent of Fox Trotters are used for pleasure riding purposes. Although the breed's characteristics have been in formulation for generations, the registry was not formed until 1948. Originally, gaited breeds such as the Standardbred and the Tennessee Walking Horse were crossed with such traditional breeds as the Morgan, Arabian, and Thoroughbred to achieve the desired characteristics. The stud book was closed in 1982 to any further outside influence, and now all registered Fox Trotters must be the product of a Fox Trotter and Fox Trotter cross. The breed's gaits include the flat-footed walk, the fox trot, and the rocking chair canter.

Colors: All colors except Appaloosa patterns
Height: 14 to 16hh
Type: Light horse
Gaited: Yes
ALBC Status: None
RBC Status: None
RBST Status: None
Breed Organization Website: www.mfthba.com or call 417-683-2468

Morab

Breed Description: Thanks to William Randolph Hearst, this breed of Morgan-to-Arabian origin has a name. Credit for the name *Morab* is given to Hearst, who raised this beautiful breed on his ranch in California during the 1920s. Although the first Morgan/Arabian crosses occurred in the late 1800s, there was no registry for the breed until the 1970s. Prior to that time, the Morabs were registered in the Morgan stud book. The registry indicates that the ideal Morab is a balanced mixture of muscle and refinement, which is not surprising when you consider the strengths of the foundation breeds. While the name implies that the Morab is a first-generation cross between a Morgan and an Arabian, in actuality, the Morab's pedigree can be comprised of up to 75 percent of either breed.

Colors: All colors. Pinto patterns are excluded, although sabino characteristics are allowed.
Height: 14.1 to 15.3hh
Type: Light horse
Gaited: No
ALBC Status: None
RBC Status: None
RBST Status: None
Breed Organization Website: www.morab.com

Morgan

Breed Description: The Morgan is a truly beautiful American horse breed that descended from a stallion who single-handedly spearheaded the entire breed in the early 1800s. The name of this famous stallion was Figure, but his name was later changed to Justin Morgan in honor of his owner. Over time, the Justin Morgan horses became known as Morgans. The background of this incredibly prepotent stallion remains a mystery even today, although speculation abounds. Some believe that Justin Morgan was a son of the Thoroughbred stallion True Briton, while others believe this is impossible because Justin Morgan and his offspring bear little resemblance to Thoroughbreds. Other sources believe that there could be Welsh Cob or Friesian breeding behind Justin Morgan, or put forth the possibility that he could have been a Cheval Canadian. While we may never know the absolute truth behind Justin Morgan's mysterious past, the fact remains that the Morgan breed of today exists because of this special stallion. According to the American Morgan Horse Association, there are seven important characteristics of Morgan horses: animation, adaptability, stamina, attitude, vigor, tractability, and alertness. Trivia tidbit: Laura Ingalls Wilder (author of the classic Little House books) raised Morgan horses in Missouri with her husband, Almanzo, during the early twentieth century. Almanzo's parents were Morgan breeders in New York State during the mid-1800s. At that time, the price of a quality Morgan two-year-old was $200.

Colors: Chestnut, black, bay, brown, palomino, buckskin, smoky black, dun, cremello, perlino, smoky cream, silver dapple, grey, and roan. The sabino, splashed white, and frame overo patterns are present within the Morgan breed, although the tobiano pattern is not.
Height: 14 to 15.2hh
Type: Light horse
Gaited: Occasionally
ALBC Status: None
RBC Status: At risk
RBST Status: None
Breed Organization Website: www.morganhorse.com or call 802-985-4944

Mountain Pleasure Horse

Breed Description: Easily confused with the more commonly recognized Rocky Mountain Horse, the Mountain Pleasure Horse is actually a historic breed that predates the Rocky Mountain Horse by over 100 years. Some sources suggest that Mountain Pleasure Horses were the original stock from which some of the other gaited breeds developed, including the Tennessee Walking Horse. In the case of its own history, the Mountain Pleasure Horse is said to have descended from Spanish horses that existed in Kentucky and Tennessee in the 1830s and 1840s. The significant difference between the Mountain Pleasure Horse and the Rocky Mountain Horse is the fact that Old Tobe, the foundation stallion of the Rocky Mountain Horse, has had very little impact on the Mountain Pleasure Horse breed. Trivia tidbit: Some people believe that Roy Rogers' horse, Trigger, was a Mountain Pleasure Horse.

Colors: Commonly found in a variety of solid colors, also grey and roan. The chocolate chestnut commonly found in the Rocky Mountain Horse is seen less often in the Mountain Pleasure Horse.
Height: 14.2 to 15.2hh
Type: Light horse
Gaited: Yes
ALBC Status: Watch
RBC Status: None
RBST Status: None
Breed Organization Website: www.mountainpleasurehorse.org

Mule

Breed Description: In its most basic description, a mule is the product of a cross between a male Donkey and a female horse. The outcome of the combination of Donkey and horse characteristics results in the fact that a mule exhibits traits of both the Donkey and the horse. The mule displays the long ears for which the Donkey is famous, but the mule also is more horse-like in its physical conformation. Mules can be found in a variety of sizes, from the smallest of Miniature Mules to the impressive Draft Mule (the product of a Mammoth Jack Donkey and a draft horse mare). Mules are nearly always sterile, and it is only the rare female mule (sometimes called a Molly Mule) that is able to produce a foal. Some estimate the odds of a female mule producing a foal to be as rare as one in a million. Mules have long been used as pack or draft animals, and many people find their unique appearance and personality to be quite endearing, which is entirely the opposite of the "stubborn as a mule" stereotype that is unfortunately connected with mules.

Colors: All, except for pinto patterns. Appaloosa patterns are regularly seen.
Height: A wide range, from 36 inches to 17hh, depending on the parentage
Type: Mule
Gaited: No
ALBC Status: None
RBC Status: None
RBST Status: None
Breed Organization Website: www.lovelongears.com

National Spotted Saddle Horse

Breed Description: In 1979, the National Spotted Saddle Horse Association (NSSHA) was established to offer a registry for gaited horses that exhibit pinto patterns. Horses of any breed may be registered; however, the individual animal must meet eligibility requirements. The horse must display tobiano, sabino, overo, or tovero characteristics, and the horse must also exhibit gaited characteristics including a flat walk, a running walk, or a pace rack. The ideal National Spotted Saddle Horse resembles a Tennessee Walking Horse in appearance, although the NSSH is usually somewhat smaller in size and more substantially built.

Colors: Tobiano, sabino, overo, or tovero
Height: 13.3 to 15.2hh
Type: Light horse
Gaited: Yes
ALBC Status: None
RBC Status: None
RBST Status: None
Breed Organization Website: www.nssha.com

New Forest Pony

Breed Description: The New Forest Pony is named for the New Forest of England from which it originates. The Society for Improvement of the New Forest Pony was founded in 1891. Breeders sought to improve the quality of the native ponies by selectively crossing them with Welsh Mountain Ponies, Fell Ponies, and Dartmoor Ponies, along with Thoroughbreds and Arabians. The New Forest Pony breed standard calls for the basic qualities that are typical of pony breeds, including a pony head, good bone, nice depth of body, and quality limbs and feet. As with many of the other native pony breeds, certain individual New Forest Ponies can be slighter and more refined than the breed standard ideal. They are talented ponies that are well suited for working under saddle.

Colors: All colors, except for cremello. Pinto and Appaloosa markings are excluded. White markings in general are discouraged.
Height: 12 to 14hh; 14.2hh is the upper limit (there is no lower limit)
Type: Pony
Gaited: No
ALBC Status: None
RBC Status: None
RBST Status: Other
Breed Organization Website: www.newforestponysocietyna.org or www.newforestpony.net

North American Sportpony

Breed Description: The North American Sportpony Registry (NASPR) began in 1997 in order to offer a place for American-bred, Warmblood-type ponies to be registered and inspected. Initially called the American Sport Pony Registry, the name was changed in 2008 to reflect the inclusion of Canadian-bred Sportponies. The fundamental characteristics of the NASPR ponies are athleticism and talent. Breed type leans toward a more horse-like appearance as opposed to the stereotypical "fat shaggy pony" appearance that is often associated with ponies in general. Several inspections are held throughout North America each year where ponies can be evaluated in comparison to the NASPR standard.

Colors: All
Height: 13.2 to 14.2hh, although breeding stock often falls outside of these parameters
Type: Pony
Gaited: No
ALBC Status: None
RBC Status: None
RBST Status: None
Breed Organization Website: www.americansportpony.com

North American Spotted Draft

Breed Description: The North American Spotted Draft is the second of two draft breeds that have been developed in the United States (the first was the American Cream Draft), and the North American Spotted Draft registry was formed in 1995. Subsequently, more than 3,000 Spotted Drafts have been registered. There is less uniformity in breed type than you might see in some of the other draft breeds, but this is because the North American Spotted Draft places a high emphasis on the breed's pinto coloring; therefore, the breed's type can vary from Percheron to Belgian to Clydesdale, etc., depending on the background of each individual animal.

Colors: Any base color, accompanied by tobiano, overo, or tovero color patterns
Height: 16 to 17hh
Type: Draft horse
Gaited: No
ALBC Status: None
RBC Status: None
RBST Status: None
Breed Organization Website: www.nasdha.net

Norwegian Fjord

Breed Description: Used by the Vikings and later as a farm horse, the Norwegian Fjord is an ancient breed with a long history of dedicated and devoted service. Today this substantial equine is very popular as a driving animal. Primitive markings and dun coloring are the hallmarks of this distinctive breed, which is always a form of dun. Also unique to the Norwegian Fjord is its two-toned mane of lighter hair with a dark stripe down the middle. Fjord enthusiasts typically keep the manes cut short, which accentuates the striking appearance of the bi-colored mane. The Norwegian Fjord is notably long-lived and is popular as a family horse, given its gentle, sensible disposition. Trivia tidbit: The black stripe in the Fjord's mane is called the *midtstol*, and the black stripe in its tail is called the *halefjaer*.

Colors: Always a form of dun, including grey dun (*gråblakk*), brown dun (*brunblakk*), red dun (*rodblakk*), white dun (*ulsblakk*), or the very rare yellow dun (*gulblakk*)
Height: 13.2 to 14.2hh
Type: Light horse, despite its pony size
Gaited: No
ALBC Status: None
RBC Status: None
RBST Status: None
Breed Organization Website: www.nfhr.org

Oldenburg

Breed Description: The Oldenburg is a Warmblood breed that has evolved from a wide array of breeds, each lending something special to this admired and extremely talented breed. Thoroughbreds, Selle Francais, and Anglo-Arabians all contributed to the Oldenburg's foundation, as did several other Warmblood breeds from Holsteiners to Trakehners. While the Oldenburg breed has been less stringent in its registration requirements than some of the other Warmblood breeds, it has been enormously successful in the top levels of equestrian sport. Trivia tidbit: Anky van Grunsven, four-time Olympic medalist in dressage, rode an Oldenburg named Bonfire.

Colors: Commonly black, bay, or grey, but other colors are acceptable
Height: Over 16hh
Type: Light horse, Warmblood
Gaited: No
ALBC Status: None
RBC Status: None
RBST Status: None
Breed Organization Website: www.oldenburghorse.com

Palomino

Breed Description: While the Palomino is not technically considered a breed, the Palomino Horse Breeders Association (PHBA) was established in 1941 to register equines of palomino coloring as long as they satisfy other selected criteria of size, breed, and color. For years, the genetic inheritance of palomino coloring was poorly understood, and some people believed that palomino horses bred to other palominos should result in 100 percent palomino offspring. Unfortunately for palomino breeders, this is not the case. Palomino is produced by the action of a cream gene that dilutes a chestnut horse to palomino. Every palomino horse has a 50 percent chance of passing its cream gene to its offspring. In the case of breeding two palominos, their offspring have a 50 percent chance of being palomino, a 25 percent chance of being chestnut (if neither parent passed their cream gene), and a 25 percent chance of being cremello. The cremello coloring occurs when a horse inherits two copies of the cream gene. When a cremello is crossed with a chestnut, the result is 100 percent palomino; however, palomino breeders tend to avoid this type of cross, as it is felt that the cremello/chestnut crosses do not produce the best shades of palomino. The PHBA prohibits the registration of palominos that exhibit any type of dorsal stripe, zebra markings, or pinto or Appaloosa characteristics. In addition, cremellos or perlinos are ineligible for registration, as are grey horses that are born palomino and palomino roans. Breeds that are recognized by the PHBA include American Quarter Horse, American Paint Horse, Appaloosa, American Saddlebred, Morgan, Arabian, Thoroughbred, Tennessee Walking Horse, Morab, Mountain Pleasure Horse, Rocky Mountain Horse, Holsteiner, Pinto, Half-Arabian, Missouri Fox Trotter, and Quarab.

Colors: Palomino; ideally the golden shade of a newly minted gold coin, accompanied by a white mane and tail
Height: 14 to 17hh
Type: Light horse, colored breed. Horses exhibiting draft or pony characteristics are ineligible for registration.
Gaited: Varies
ALBC Status: None
RBC Status: None
RBST Status: None
Breed Organization Website: www.palominohba.com

Paso Fino

Breed Description: Known as *Los Caballos de Paso Fino* in Spanish, the Paso Fino is of Puerto Rican origin. Its name translates to "the horses with the fine walk." The breed was developed with considerable Spanish horse influence, including the Andalusian and the Spanish Jennet. The Paso Fino is a gaited breed, and its three gaits include the *paso fino* (walk), *paso corto* (trot), and the *paso largo* (canter). Originally brought to the United States in the mid-twentieth century, the Paso Fino has gained a reputation as a splendid trail and pleasure horse.

Colors: All
Height: 14 to 15hh
Type: Light horse
Gaited: Yes
ALBC Status: None
RBC Status: None
RBST Status: None
Breed Organization Website: www.pfha.org

Percheron

Breed Description: Named for the Le Perche area of France, the impressive Percheron breed is unique among draft breeds in that it is commonly grey. The grey coloring is attributed to the foundation breeds of the Percheron—the Arabian and native French horses. In the 1944 book *Horses of Britain*, author Lady Wentworth expressed the opinion that grey "is the only proper color for a Percheron. Black is undesirable." Because Wentworth was a renowned Arabian breeder, it is not surprising that she would assume that the grey Percherons were of higher quality, as they would have apparently been more influenced by the Arabian than the black Percherons. The Arabian influence not only affected the color of the Percheron, it also significantly affected the Percheron's head and movement. The result has been that the Percheron has a longer stride than some of the other draft breeds, and their heads are also inclined to be more attractive. The Percheron stud book was opened in 1893, and today the breed is very popular in the United States, where they are commonly used in draft hitches and are popular driving animals.

Colors: Commonly grey and black
Height: 16 to 18hh
Type: Draft horse
Gaited: No
ALBC Status: Recovering
RBC Status: None
RBST Status: None
Breed Organization Website: www.percheronhorse.org

Peruvian Paso

Breed Description: Although you might assume otherwise, the Peruvian Paso and the Paso Fino are actually two distinctly different breeds, having been developed in completely different places. The Paso Fino is a Puerto Rican breed, while the Peruvian Paso originated in Peru. The Spanish conquistadors in Peru possessed many Spanish horses, including Andalusian, Barb, and Spanish Jennet, and it was a combination of these breeds that proved to be the foundation of the Peruvian Paso. Some sources suggest that Friesian blood was also infused. The first Peruvian Pasos were brought to the United States during the 1960s, and they are noted for their incredibly smooth gaits, which make them very popular with pleasure and trail riders. The Peruvian Paso is admired for its unique *paso llano* gait, which is a four-beat lateral gait exclusive to the breed. The *sobreandando* gait is also associated with this breed. *Termino* is a word that you will often hear associated with Peruvian Pasos, and it refers to a particular type of action, characterized by an extended arch of the horse's front legs. *Brio* is another Peruvian Paso term that refers to the inner presence and fire that the Peruvian Paso exhibits.

Colors: All colors, including black, brown, chestnut, bay, grey, palomino, buckskin, roan, dun, and grulla
Height: 14.1 to 15.2hh
Type: Light horse
Gaited: Yes
ALBC Status: None
RBC Status: None
RBST Status: None
Breed Organization Website: www.napha.net

Pintabian

Breed Description: For Arabian enthusiasts who admire the tobiano color pattern, the Pintabian breed is undoubtedly a dream come true! The Pintabian exhibits the fundamental traits that define the Arabian: beauty, elegance, refinement, and stamina, with the added bonus of pinto coloring. The Pintabian Horse Registry, Inc. (PHRI), founded in 1992, recognizes the tobiano color pattern in Pintabians, although Pintabians that do not exhibit the proper color pattern can be registered in the Breeding Stock division of the registry. For registration in the Colored division, the Pintabian must exhibit the tobiano pattern, as well as meet the registration criteria of being more than 99 percent Arabian, although not 100 percent.

Colors: Must exhibit the tobiano pattern
Height: 14.2 to 15.2hh
Type: Light horse
Gaited: No
ALBC Status: None
RBC Status: None
RBST Status: None
Breed Organization Website: www.pintabianregistry.com

Pinto

Breed Description: Unlike the American Paint Horse Association (APHA), which restricts registration to Quarter Horses, Thoroughbreds, and Paint Horses, the American Pinto Horse Association (PtHA) registers equines of many breeds. These include the American Paint Horse, American Quarter Horse, Oldenburg, Trakehner, Hanoverian, Holsteiner, Thoroughbred, Andalusian/Lusitano, Tennessee Walking Horse, National Show Horse, American Saddlebred, American Standardbred, Westphalian, Morgan, Missouri Fox Trotting Horse, Arabian, Hackney, Connemara, American Quarter Pony, Shetland, Welsh, and Miniature Horses. To be registered with the PtHA, a horse must meet certain color criteria (the horse must exhibit the tobiano or overo pattern and exhibit a particular amount of white on certain areas of the body) and must not be an Appaloosa, Mule, or draft breed. The PtHA recognizes four different types in their registry: hunter, stock, pleasure, and saddle. The PtHA is not as large as the APHA, and has fewer than 5,000 pinto foals registered each year. Trivia tidbit: *Bonanza* fans will undoubtedly remember a brightly marked Pinto horse named Cochise that belonged to Little Joe Cartwright.

Colors: Tobiano and overo patterns
Height: Horses: 14hh and over
  Ponies: 38 to 55 inches
  Miniature: under 34 inches
  Miniature B: 34 to 38 inches
Type: Varies
Gaited: Occasionally
ALBC Status: None
RBC Status: None
RBST Status: None
Breed Organization Website: www.ptha.org

Polo Pony

Breed Description: As with hunter ponies, Palominos, and Buckskins, the Polo Pony is not considered to be a breed, but rather a type of light horse that is used for the game of polo. The Polo Pony must exhibit several attributes that are necessary in order to excel in the game of polo, and these characteristics include speed, agility, bravery, and stamina. Polo Ponies are often produced through crossbreeding, especially with British native ponies, Thoroughbreds, Quarter Horses, and Criollos (a native Argentinian breed).

Colors: All
Height: Despite its name, the Polo Pony is typically larger than pony size, averaging 15.1hh
Type: Light horse
Gaited: No
ALBC Status: None
RBC Status: None
RBST Status: None
Breed Organization Website: www.americanpolohorse.com

Pony of the Americas (POA)

Breed Description: This popular American pony breed originated in Iowa in the 1950s when the future foundation sire of the breed was born. Black Hand I was a Shetland/Appaloosa/Arabian crossbred. His owner was quite pleased with the unique qualities of this young colt that combined the smaller size of the pony with the eye-catching color pattern of the Appaloosa. Over the years, other infusions into the Pony of the Americas (POA) breed included the Quarter Horse, the Welsh Pony, the Morgan, and the Thoroughbred. The POA is a stock-type pony that has become a very popular choice as a children's pony. For registration, the pony's spots must be visible at a distance of 40 feet. The registry also maintains a hardship division so that animals fulfilling the physical requirements can be granted registration papers.

Colors: The Pony of the Americas Club, Inc., recognizes seven patterns: snowflake, frost, blanket, leopard, white with dark spots, marbleized roan, and few-spot leopard
Height: 46 to 56 inches
Type: Pony
Gaited: No
ALBC Status: None
RBC Status: None
RBST Status: None
Breed Organization Website: www.poac.org or call 317-788-0107

Rocky Mountain Horse

Breed Description: Although its name implies differently, the Rocky Mountain Horse does not hail from the Rocky Mountains, but rather from Kentucky. The foundation sire of this popular gaited breed was Old Tobe, who was brought to Kentucky from the Rocky Mountains in the middle of the twentieth century. He was locally known as the Rocky Mountain Horse, and the breed owes its name to him. The Rocky Mountain Horse Association was founded in 1986, and the breed is popular with trail and pleasure riders who are fond of the breed's four-beat, single-foot gait. The breed's signature color is a chocolate-chestnut color, accompanied by a flaxen mane and tail.

Colors: All solid colors, but ideally a chocolate-chestnut color with a flaxen mane and tail. Conservative white markings are allowed.
Height: 14.2 to 16hh
Type: Light horse
Gaited: Yes
ALBC Status: Watch
RBC Status: None
RBST Status: None
Breed Organization Website: www.rmhorse.com

Selle Francais

Breed Description: Known as *Le Cheval de Salle Francais* in France, the name *Selle Francais* roughly translates to "the French Saddle Horse." The Selle Francais breed is noted for its talented show jumpers, as well as for the breed's achievement in dressage. They are notably easy-going and trainable, with an elegance that stems from the foundation breeds that include Arabian, Anglo-Arabian, and Thoroughbred. Other foundation breeds of the Selle Francais included the native French (Norman) horses and the French Trotters.

Colors: Commonly chestnut, but also bay, roan, or grey
Height: 15.1 to 17hh
Type: Light horse
Gaited: No
ALBC Status: None
RBC Status: None
RBST Status: None
Breed Organization Website: www.fnsf.eu

Shagya Arabian

Breed Description: A grey Arabian stallion named Shagya was imported from Syria to Hungary in the 1830s. It was his influence that eventually resulted in today's Shagya Arabian breed. Taller and well-suited for sport-horse disciplines, the Shagya Arabian is not considered to be a purebred Arabian because of the occasional infusion of non-Arabian breeding that was utilized during the breed's development. The Shagya Arabian breed has, however, maintained its own stud book that is now closed to outside influence, and the Shagya Arabian breed itself is considered purebred. The breed was rare in the United States until the 1980s, at which time the North American Shagya Arabian Society (NASS) was formed. The NASS hosts an annual inspection tour for the Shagya breed, similar to the sport-horse inspections that are held for many of the Warmblood breeds.

Colors: Commonly grey; also black, bay, and chestnut
Height: 15 to 16hh
Type: Light horse
Gaited: No
ALBC Status: None
RBC Status: None
RBST Status: None
Breed Organization Website: www.shagya.net

Shetland (Classic)

Breed Description: Originating from the Scottish Shetland Islands, the classic Shetland Pony has a long history of life on the harsh and rugged island terrain. In the 1800s, British coal miners realized the value of the Shetland Pony for use in the coal mines, and as a result, many Shetland Ponies went to work as pit ponies. Their strength and stamina was impressive, despite their minute size. When ponies were no longer needed for mining work, their popularity as children's ponies increased. Today's classic Shetland Pony is prized for its traditional substance and native pony character.

Colors: All colors except "spotted." According to Lady Wentworth's 1944 book, *Horses of Britain*, "Piebald and skewbald are prehistoric original colors [of Shetlands] and should be preserved and encouraged."
Height: Up to 46 inches at the withers
Type: Pony
Gaited: No
ALBC Status: None
RBC Status: Critical
RBST Status: Other
Breed Organization Website: www.shetlandminiature.com or call 309-263-4044

Shetland (Modern)

Breed Description: Although they share the same name, the Modern Shetland Pony differs considerably in appearance from the Classic Shetland Pony. This is because the Shetland Ponies were crossbred with other breeds such as Hackney and Welsh in an attempt to accentuate the action and increase the size of the pony. However, according to pony expert Lorna Howlett's book, *The Complete Book of Ponies*, "The Shetland Pony, bred in strict accordance to its Standard of Excellence, is far removed from the elegant, Hackney-type, high-stepping animal depicted on the letter-head of the American Shetland Pony Club." Although these words are strong, there are many Shetland Pony enthusiasts who share the same sentiments, both from the perspective of the Classic Shetland lovers who wish to see the original breed standard preserved, and from the perspective of the Modern Shetland enthusiasts who are pleased with their taller, more Hackney-like version of the Shetland Pony.

Colors: All, except spotted
Height: There are two types: 42 inches and under; and 43 to 46 inches, but not over 46 inches
Type: Pony
Gaited: No
ALBC Status: None
RBC Status: None
RBST Status: None
Breed Organization Website: www.shetlandminiature.com or call 309-263-4044

Shire

Breed Description: According to *Horses of Britain*, the Shire breed represents "the ancient, prehistoric Great Horse of Europe, though they have been strongly influenced by the Spanish type of battle horse which had the same white legs, the blaze, the crested neck, and the profuse mane." The Shire is named for the English shires from whence it originates and is notable for its height and substance. The Shire Horse Society was formed in the United

Kingdom in 1884. The official breed standard calls for a Shire stallion to "have good feet and joints; the feet should be wide and big around the top of the coronets with sufficient length in the pasterns. When in motion, he should go with force using both knees and hocks, which latter should be kept close together. He should go straight and true before and behind. A good Shire stallion should have strong character." Trivia tidbit: The tallest horse ever recorded was a Shire that stood 21.2½hh.

Colors: Bay, brown, black, and grey. Chestnut is rare, but accepted in the American Shire Horse Society. In the United Kingdom, chestnut or roan stallions are not acceptable. Roan is allowed for mares and geldings in the United Kingdom. White markings are very common, although not particularly desirable. White patches on the body are discouraged.

Height: Over 16.2hh

Type: Draft

Gaited: No

ALBC Status: Watch

RBC Status: Endangered

RBST Status: At Risk

Breed Organization Website: www.shirehorse.org

Suffolk Punch

Breed Description: As with the Morgan horse, the Suffolk Punch breed was indelibly affected by a single stallion. In the case of the Suffolk Punch, the stallion was Thomas Crisp's Horse of Ufford in the 1700s, who is said to be the ancestor of every Suffolk Punch horse in existence. Unlike some of the other British draft breeds that are noted for their extreme size and heavy feathering, the Suffolk Punch is a shorter, more compact breed. The breed standard calls for "long clean hocks on short cannon bones free from coarse hair." They have a long history as farm horses and are well suited for harness work, with the quiet temperament necessary for a work horse.

Colors: Chestnut (always spelled "chesnut" without the 't' in the Suffolk Punch breed). White markings are unusual, except for a small star or stripe on the face.
Height: 16.1 to 17.1hh
Type: Draft
Gaited: No
ALBC Status: Critical
RBC Status: Endangered
RBST Status: Critical
Breed Organization Website: www.suffolkpunch.com

Swedish Warmblood

Breed Description: The high caliber of the Swedish Warmblood breed is certainly owed in part to its long history. Developed in Sweden during the 1600s, the breed's foundation stock ranged from Friesians to Arabians and Trakehners to Spanish horses. Great thought and consideration was taken in the selective breeding of Swedish Warmbloods over the past four centuries. Today the breed's consistent type and quality makes it a popular choice in equestrian sports from driving to dressage.

Colors: All solid colors, including bay, brown, chestnut, and grey
Height: 16.2 to 17hh
Type: Light horse, Warmblood
Gaited: No
ALBC Status: None
RBC Status: None
RBST Status: None
Breed Organization Website: www.swanaoffice.org

Tennessee Walking Horse

Breed Description: The Tennessee Walking Horse is well-known for its three distinctive gaits (the flat-footed walk, the running walk, and the rocking chair canter), and is greatly admired for its smooth and comfortable ride, making it a popular choice as a trail mount. The breed was developed in Tennessee, and its foundation stallion was a Standardbred/Morgan cross named Black Allan. It is interesting to note that Black Allan was sired by a descendent of Rysdyk's Hambletonian and also had Justin Morgan in his pedigree. Other background breeds of the Tennessee Walking Horse include the Saddlebred, Narragansett Pacer, and Thoroughbred, along with some Spanish influence. Trivia tidbit: Trigger, Jr., one of Roy Rogers' mounts, was a palomino Tennessee Walking Horse.

Colors: Black, smoky black, smoky cream, bay, brown, buckskin, perlino, classic champagne, classic cream champagne, gold champagne, gold cream champagne, amber champagne, amber cream, chestnut/sorrel, palomino, and cremello. The Tennessee Walking Horse Breeders' and Exhibitors' Assocation also recognizes sabino, tobiano, overo, and tobiano/sabino color patterns, as well as four color-modifying genes: roan, grey, dun, and silver.
Height: 15 to 16hh
Type: Light horse
Gaited: Yes
ALBC Status: None
RBC Status: None
RBST Status: None
Breed Organization Website: www.twhbea.com or call 931-359-1574

Thoroughbred

Breed Description: It is difficult to know where to start when attempting to detail the extensive history, the immense influence, and the vast importance of the Thoroughbred horse. Its worldwide fame as a racing breed is well-known, as is its far-reaching influence on many other breeds from the Quarter Horse to the Dutch Warmblood. The Thoroughbred is an English breed that was developed during the eighteenth century, produced from three primary foundation sires: the Darley Arabian, the Godolphin Arabian, and the Byerley Turk. It has been said that all Thoroughbreds today descend from one or more of these stallions. While the breed is obviously best known for its racing ability, Thoroughbreds are also well-known for their suitability as hunters, jumpers, and eventers, as well as for their talent in dressage. Lady Wentworth probably best summed up the breed in her 1944 book, *Horses of Britain*, when she wrote, "A Thoroughbred should have all the best points of a technically good horse with an added look of quality and aristocratic blood." Trivia tidbit: The ultimate accolade in American Thoroughbred racing is the Triple Crown, which is awarded to Thoroughbreds that win three important races (the Kentucky Derby, the Preakness Stakes, and the Belmont Stakes) in the same year. As of 2008, only eleven horses have ever won the Triple Crown in over a century of racing. These famous individuals include Sir Barton, Gallant Fox, Omaha, War Admiral, Whirlaway, Count Fleet, Assault, Citation, Secretariat, Seattle Slew, and Affirmed. There has not been a Triple Crown winner since 1978.

Colors: Black, bay, chestnut, brown, and grey
Height: 15.2 to 17hh
Type: Light horse
Gaited: No
ALBC Status: None
RBC Status: None
RBST Status: None
Breed Organization Website: www.jockeyclub.com

Trakehner

Breed Description: Named for the original breeding farm at Trakehnen in the former East Prussia, the Trakehner breed is somewhat more refined than other Warmbloods. The breed derived from native Schweiken horses that were selectively crossed with Thoroughbred and Arabian horses. The ravages of World War II resulted in the loss of thousands of Trakehners, and only a few hundred remained in East Prussia after the war ended. Thankfully, the breed has recovered itself and is now flourishing. According to the breeding goal established by the Trakehner Verband bylaws, "A good character, a well-balanced temperament, intelligence, willingness to work, as well as endurance and hardiness during work are to be particularly apparent characteristics" in the Trakehner breed.

Colors: All
Height: 15.1 to 16.3hh (stallions must be a minimum of 15.3hh)
Type: Light horse, Warmblood
Gaited: No
ALBC Status: None
RBC Status: None
RBST Status: None
Breed Organization Website: www.americantrakehner.com

Welara

Breed Description: "The best ponies in the world can be produced from Arabs and Welsh crosses," wrote Lady Wentworth in her 1951 book, *Horses in the Making.* Wentworth, a renowned breeder of both Arabians and Welsh, is credited with the idea behind the Welara, an Arabian/Welsh crossbreed suitable for a child's pony. The ideal outcome was to increase the size of the Welsh Pony while still preserving the beauty, movement, and quality of both breeds. Interest in the Welara is beginning to increase in the United States, where the American Welara Pony Registry was formed in 1981. As long as the Welara is registered in the Half-Welsh or Half-Arabian registries of those respective breeds, the Welara can also compete in Arabian- and Welsh-sanctioned shows.

Colors: All
Height: 46 to 60 inches is ideal
Type: Pony
Gaited: No
ALBC Status: None
RBC Status: None
RBST Status: None
Breed Organization Website: www.welararegistry.com

Welsh Cob (Section D)

Breed Description: Extremely popular in the United Kingdom, the Welsh Cob is still a relative newcomer to American shores. The Welsh Pony Society of America did not change its name to the Welsh Pony and Cob Society of America until 1980, at which time registrations for Sections C and D of the stud book opened in this country. Since then, imports of Welsh Cob foundation stock from Wales to the United States have increased. Approximately forty purebred Section D foals were born in America in 2005 (the most recent year for which data is available). The official Welsh Cob breed description describes the Section D as "strong, hardy, and active, with pony character and as much substance as possible." Welsh Cobs have incredible strength and stamina and are popular for eventing, dressage, and driving.

Colors: All colors; commonly bay, chestnut, and black; also palomino, cremello, buckskin, perlino, and smoky black. Rarely grey and roan. Extensive sabino and splash patterns are accepted for registration in the United States, but the breed standard discourages them. Appaloosa patterns are not accepted.

Height: Over 13.2hh with no upper height limit. Section D Welsh Cobs that are less than 13.2hh are re-registered in Section C of the stud book.

Type: Pony

Gaited: No

ALBC Status: None

RBC Status: Vulnerable

RBST Status: Other

Breed Organization Website: www.welshpony.org

Welsh Mountain Pony (Section A)

Breed Description: One of the best descriptions of the Welsh Mountain Pony appears in the very first stud book of the Welsh Pony Society of America (now the Welsh Pony and Cob Society of America), published in 1911. It reads "Welsh ponies . . . are noted for their soundness in limb and wind, as well as for their vigor and hardy constitutions. These traits are the result of environment; bred and reared in the mountains, galloping down the steep sides; jumping across ravines from crag to crag as they do with the agility of rabbits." The Section A Welsh Mountain Pony is generally considered to be the most beautiful pony breed in the world, a statement easily confirmed by one glance at the Welsh Mountain Pony's most distinctive characteristic—its small, clean-cut head, accentuated by huge, dark eyes, and tiny, pointed ears. The Welsh Mountain Pony's action is also particularly impressive with its free reaching stride with excellent use of the knees and hocks. Section As have long been favored as children's mounts, although they are also quite popular as riding mounts for small adults, and their suitability for harness is also well-documented. Lieutenant Colonel Sir Harry Llewellyn, a noted Welsh show judge, aptly summarized the Welsh Mountain Pony in two words: "Robust quality."

Colors: All colors; commonly grey and chestnut; also bay, black, palomino, cremello, buckskin, perlino, smoky black, and roan. Extensive sabino and splash patterns are accepted for registration in the United States, but the breed standard discourages them. Appaloosa patterns are not accepted.

Height: The upper height limit for Welsh Mountain Ponies (Section A) in the United States and Canada is 12.2hh. The upper height limit is 12hh in the United Kingdom and Australia.

Type: Pony

Gaited: No

ALBC Status: None

RBC Status: Vulnerable

RBST Status: Vulnerable

Breed Organization Website: www.welshpony.org

Welsh Pony (Section B)

Breed Description: In the 1930s, Welsh Pony enthusiasts in Wales began discussing the need for a pony with "riding quality" and lower action than the Section A, a Welsh Pony that would be perfectly suitable as a child's mount after outgrowing their Section A pony. And it was thus that the Section B Welsh Pony was established, filling a niche that had been previously unoccupied within the Welsh Pony and Cob Society. Some of the original Section Bs contained a bit of Arabian or Barb blood, but the stud book was quickly closed to further outside influence during the 1950s. The Section Bs of today are extremely popular as children's ponies, often competing very successfully in hunter classes as well as doing exceptionally well in flat classes. Section Bs are a marvelous combination of the best of native pony breeding in a slightly larger package.

Colors: All colors; commonly grey and chestnut; also bay, black, palomino, cremello, buckskin, perlino, smoky black, and roan. Extensive sabino and splash patterns are accepted for registration in the United States, but the breed standard discourages them. Appaloosa patterns are not accepted.

Height: The upper height limit for Section Bs varies worldwide; in the United States, they may be up to 14.2hh, in Canada they may be up to 14hh, and they may be up to 13.2hh in the United Kingdom. There is no lower height limit, although Section Bs are rarely less than 12hh. Many measure in at about 12.2hh.

Type: Pony

Gaited: No

ALBC Status: None

RBC Status: Vulnerable

RBST Status: Other

Breed Organization Website: www.welshpony.org

Welsh Pony of Cob Type (Section C)

Breed Description: Although it is currently the rarest of the Welsh breeds, the Welsh Pony of Cob Type (Section C) has been steadily gaining in popularity and recognition in both the United Kingdom and the United States. Although the Section C is approximately the same height as the Section B Welsh Pony, the two are significantly different in terms of type. The Section C is best described as a larger version of the Welsh Mountain Pony, a solid individual with all of the characteristic native pony quality, coupled with the larger size and scope of the Welsh Cob. A huge, ground-covering stride and ample bone and substance make the Section C a perfect "all-around" pony. The Section C was originally produced by crossing Section A Welsh Mountain Ponies with Section D Welsh Cobs, and this foundation cross is still practiced today. Many breeding farms, however, specialize in Section C bloodlines and perpetuate their breeding programs by crossing Section Cs only with other Section Cs. Section Bs are not typically used in the production of Section Cs, as many breeders feel that Section B crosses are not in the best interests of the Section C.

Colors: All colors; commonly bay, chestnut, and black; also palomino, cremello, buckskin, perlino, and smoky black. Rarely grey and roan. Extensive sabino and splash patterns are accepted for registration in the United States, but the breed standard discourages them. Appaloosa patterns are not accepted.

Height: Up to 13.2hh. There is no lower limit, but the Section C is usually at least 12hh. Section Cs that grow over 13.2hh are re-registered as Section Ds in the stud book.

Type: Pony

Gaited: No

ALBC Status: None

RBC Status: Vulnerable

RBST Status: Other

Breed Organization Website: www.welshpony.org

Glossary

Agouti gene: The gene responsible for the restriction of black to the points, as seen in colors such as bay or buckskin. Recessive agouti genes do not restrict black to the points, resulting in a uniformly black animal (such as a black, black roan, or smoky black).

Allele: The term used to describe a pair of genes.

Breed standard: The list of a breed's ideal characteristics, as outlined by a breed registry or organization.

Colt: A male horse under the age of four.

Concave profile: Also known as a dished head, it is a head that gradually tapers to the muzzle, commonly seen in Arabians and Welsh Mountain Ponies.

Conformation: The structure of a horse's body; the way a horse is "built."

Convex profile: Also known as a Roman nose, it is a head that gradually tapers outward, commonly seen in some draft horse breeds.

Cream gene: See dilute gene.

Crossbred: Not a purebred. The term crossbred typically refers to a first generation cross between two purebreds, such as a Welsh/Thoroughbred cross or an Appaloosa/Arabian cross.

Dam: The mother of a horse.

Dilute gene: Also known as the cream gene, the dilute gene lightens bay to buckskin, chestnut to palomino, and black to smoky black. It is a dominant gene.

Feather: The long hair that grows on the lower legs and fetlocks, commonly seen in native pony breeds and heavy draft breeds.

Filly: A female horse under the age of four.

Foal: A young horse, usually less than six months old.

Gaited: A horse that exhibits gaits beyond (or in place of) the basic walk, trot, and canter.

Gelding: A castrated male horse of any age.

Grade: An unregistered horse, often of unknown parentage.

Hand: The standard of measurement in horses and ponies, a hand is equal to four inches. Hands is represented as either 15 hands or its abbreviated form, 15hh.

Heterozygous: Genes are inherited in pairs called alleles. Heterozygous is the term used when the two alleles are different.

Homozygous: Homozygous is the term used when the two alleles are identical.

Mare: A female horse that is at least four years of age.

Mealy mouth: A bay or brown horse with a lighter colored muzzle.

Modifying gene: In equine color genetics, grey and roan are considered to be modifying genes because they modify the base color.

Open stud book: A breed registry that accepts horses of multiple breeds and backgrounds, regardless of pedigree.

Overo: A coat pattern that is known for producing extensively marked faces. Any pinto-marked horse that is not a distinct tobiano belongs to the overo family of coat patterns.

Points: On certain colors (bay and buckskin, for instance), the darker points include the lower legs, ear tips, mane, and tail.

Pony: An equine standing less than 14.2hh.

Prepotent: A term used for stallions that regularly and consistently produce offspring that exhibit certain characteristics, regardless of the appearance of the foal's dam.

Registered: Officially recognized by a breed registry as meeting all registration requirements and criteria.

Sabino: A coat pattern that is characterized by white markings that are speckled and flecked.

Sire: The father of a horse.

Splash: A coat pattern that is characterized by blazes that are "bottom-heavy" (wider at the muzzle than on the forehead) and extensive leg markings that lack the "speckling" that is seen in the sabino pattern.

Stallion: A male horse that is at least four years of age.

Tobiano: A coat pattern that is characterized by minimal white on the face, dark legs, and white that crosses over the back in at least one location.

Tovero: A coat pattern that exhibits characteristics of tobiano and overo.

Wall-eye: An eye that exhibits a lighter, almost white appearance.

Weanling: A foal between the ages of six months and one year.

Yearling: A young horse that has reached its first birthday but is less than two years old. (Some horse registries recognize January 1 as the birth date of all horses, thus many weanlings are considered to be yearlings as of January 1, regardless of their age.)

Index

American Bashkir Curly, 57
American Cream Draft, 58
American Paint Horse, 6, 42, 43, 49, 59
American Paint Pony, 60
American Quarter Horse, 6, 61
American Quarter Pony, 62
American Saddlebred, 16, 63
American Standardbred, 16, 64
American Warmblood, 65
Andalusian, 66
Anglo-Arabian, 67
Appaloosa, 6, 36, 40, 42, 68
Arabian, 10, 69
Belgian, 70
Black Stallion, 15
British Riding Pony, 71
Buckskin, 72
Canadian, 73
Caspian, 74
Chincoteague, 15, 75
Cleveland Bay, 18, 76
Clydesdale, 35, 77
Colonial Spanish Horse (Mustang), 78
Colorado Ranger, 50, 51, 79
colors
    amber champagne, 135
    amber cream, 135
    bay, 19, 20, 21, 23, 59, 61, 64,
      66–69, 71, 73, 74, 76, 77,
      80–83, 85–89, 92, 93, 95, 96,
      98–100, 104–106, 110, 118,
      122, 128, 129, 132, 134–136,
      139–142
    bay roan, 59, 61, 68, 89
    black, 19–21, 24, 59, 61, 64, 66, 68,
      69, 71, 73, 74, 77, 80, 81, 83–86,
      88, 89, 91–93, 95, 96, 98–100,
      103–106, 110, 118, 121, 122,
      129, 132, 135, 136, 139–142
    black bay, 66
    black roan, 19, 22
    blue roan, 19
    blue roan, 59, 61, 68, 89
    brown, 24, 59, 61, 64, 66–68, 73,
      77, 80, 81, 83, 85, 87–89, 95,
      96, 98–100, 105, 110, 122, 132,
      134–136
    buckskin roan, 19
    buckskin, 20, 22, 25, 59, 61, 66, 68,
      72, 89, 110, 122, 135, 139–142
    champagne, 25
    chestnut, 19–21, 26, 36, 40, 43, 59,
      61, 64, 66–71, 73, 74, 77, 80, 83,
      85, 88, 89, 91, 95–98, 100, 105,
      110, 122, 128, 129, 132–136,
      139–142
    chocolate-chestnut, 127
    classic champagne, 135
    classic cream champagne, 135
    cream, 58, 84
    cremello, 27, 59, 61, 66, 68, 89, 110,
      135, 139–142

dun, 27, 59, 61, 66, 68, 74, 80, 89,
    99, 104, 110, 117, 122
gold champagne, 135
gold cream champagne, 135
grey, 19–21, 28, 59, 61, 64, 67–69,
    71, 74, 78, 80, 81, 84–86, 88, 89,
    96, 98–100, 103, 105, 106, 110,
    111, 118, 121, 122, 128, 129,
    132, 134, 136, 139–142
grulla(o), 29, 59, 61, 68, 89, 122
liver chestnut, 99
palomino, 20, 22, 29, 59, 61, 66,
    68, 80, 89, 110, 119, 122, 135,
    139–142
perlino, 30, 59, 61, 66, 68, 89, 110,
    135, 139–142
red dun, 59, 61, 89
red roan, 59, 61, 68, 89
roan, 19–22, 30, 64, 66, 69, 78, 80,
    83–85, 96, 110, 111, 122, 128,
    139–142
silver dapple, 31, 110
smoky black, 31, 110, 135, 139–142
smoky cream, 110, 135
sorrel, 59, 61, 89, 135
Connemara, 80
Dales, 81
Danish Warmblood, 82
Dartmoor, 83
Donkey, 84
Dutch Warmblood, 85
Eriskay Pony, 86
Exmoor, 36, 87
Fell, 88
Fell Pony, 35
Foundation Quarter Horse, 89
French Saddle Pony, 90
Friesian, 18, 35, 36, 91
German Riding Pony, 92
Gotland, 93
Gypsy Vanner, 35, 42, 94
Hackney Horse, 95
Hackney Pony, 96
Haflinger, 34, 97
Hanoverian, 98
Highland Pony, 99
Holsteiner, 100
Hunter Pony, 101
Icelandic, 55, 102
Irish Draught, 103
Knabstrupper, 50
Lac La Croix Indian, 104
Latvian, 105
Lipizzan, 106
Miniature Horse, 14, 42, 43, 107
Missouri Fox Trotter, 42
Missouri Fox Trotting Horse, 108
Morab, 109
Morgan, 110
Mountain Pleasure Horse, 111
Mule, 112
National Spotted Saddle Horse, 113
New Forest Pony, 114

North American Sportpony, 34, 55, 115
North American Spotted Draft, 116
Norwegian Fjord, 117
Oldenburg, 118
Palomino, 119
Paso Fino, 13, 120
patterns
    Appaloosa, 50–53, 78, 108, 112
    blanket, 68, 126
    brindle, 40
    few-spot leopard, 126
    frame overo, 110
    frost, 126
    leopard, 126
    leopard complex, 52
    marbleized roan, 126
    overo, 44, 46, 49, 59, 113, 116,
      124, 135
    overo, frame, 46
    paint, 42
    piebald, 42, 43
    pinto, 36, 42, 43, 47, 75, 78, 94,
      102, 107
    sabino, 46–49, 109, 110, 113, 135,
      139–142
    sclera, 40
    skewbald, 42, 43
    snowflake, 126
    splashed white, 46, 48, 49, 110
    spotted blanket, 53
    tobiano, 19, 20, 43–46, 59, 64, 85,
      113, 116, 123, 124, 135
    tovero, 47, 49, 59, 113, 116
Percheron, 121
Peruvian Paso, 122
Pintabian, 123
Pinto, 124
Polo Pony, 125
Pony of the Americas (POA), 50, 51,
    126
Rocky Mountain Horse, 13, 127.
Selle Francais, 128
Shagya Arabian, 34, 129
Shetland, 45
    Classic, 130
    Modern, 131
    Pony, 42
Shire, 14, 35, 132
Spanish Colonial, 10
Suffolk Punch, 133
Swedish Warmblood, 134
Tennessee Walking Horse, 13, 55, 135
Thoroughbred, 6, 40, 136
Trakehner, 137
Warmblood, 34
Welara, 128
Welsch Cob, 35
    Section D, 139
Welsh Mountain Pony (Section A), 140
Welsh Pony, 11
    Section B, 141
    Section C, 142